Lecture Notes in Physics

Founding Editors

Wolf Beiglböck
Jürgen Ehlers
Klaus Hepp
Hans-Arwed Weidenmüller

Volume 1034

Series Editors

Roberta Citro, Salerno, Italy
Peter Hänggi, Augsburg, Germany
Betti Hartmann, London, UK
Morten Hjorth-Jensen, Oslo, Norway
Maciej Lewenstein, Barcelona, Spain
Satya N. Majumdar, Orsay, France
Luciano Rezzolla, Frankfurt am Main, Germany
Angel Rubio, Hamburg, Germany
Wolfgang Schleich, Ulm, Germany
Stefan Theisen, Potsdam, Germany
James D. Wells, Ann Arbor, MI, USA
Gary P. Zank, Huntsville, AL, USA

The series Lecture Notes in Physics (LNP), founded in 1969, reports new developments in physics research and teaching - quickly and informally, but with a high quality and the explicit aim to summarize and communicate current knowledge in an accessible way. Books published in this series are conceived as bridging material between advanced graduate textbooks and the forefront of research and to serve three purposes:

- to be a compact and modern up-to-date source of reference on a well-defined topic;
- to serve as an accessible introduction to the field to postgraduate students and non-specialist researchers from related areas;
- to be a source of advanced teaching material for specialized seminars, courses and schools.

Both monographs and multi-author volumes will be considered for publication. Edited volumes should however consist of a very limited number of contributions only. Proceedings will not be considered for LNP.

Volumes published in LNP are disseminated both in print and in electronic formats, the electronic archive being available at springerlink.com. The series content is indexed, abstracted and referenced by many abstracting and information services, bibliographic networks, subscription agencies, library networks, and consortia.

Proposals should be sent to a member of the Editorial Board, or directly to the responsible editor at Springer:

Dr Lisa Scalone
lisa.scalone@springernature.com

Roberto Onofrio • Luca Salasnich

Physics and Technology of Ultracold Atomic Gases

Roberto Onofrio
Physics and Astronomy
Dartmouth College
Hanover, NH, USA

Luca Salasnich
Fisica e Astronomia
Università di Padova
Padova, Italy

ISSN 0075-8450   ISSN 1616-6361   (electronic)
Lecture Notes in Physics
ISBN 978-3-031-76003-7   ISBN 978-3-031-76004-4   (eBook)
https://doi.org/10.1007/978-3-031-76004-4

© The Editor(s) (if applicable) and The Author(s), under exclusive license to Springer Nature Switzerland AG 2024

This work is subject to copyright. All rights are solely and exclusively licensed by the Publisher, whether the whole or part of the material is concerned, specifically the rights of translation, reprinting, reuse of illustrations, recitation, broadcasting, reproduction on microfilms or in any other physical way, and transmission or information storage and retrieval, electronic adaptation, computer software, or by similar or dissimilar methodology now known or hereafter developed.
The use of general descriptive names, registered names, trademarks, service marks, etc. in this publication does not imply, even in the absence of a specific statement, that such names are exempt from the relevant protective laws and regulations and therefore free for general use.
The publisher, the authors and the editors are safe to assume that the advice and information in this book are believed to be true and accurate at the date of publication. Neither the publisher nor the authors or the editors give a warranty, expressed or implied, with respect to the material contained herein or for any errors or omissions that may have been made. The publisher remains neutral with regard to jurisdictional claims in published maps and institutional affiliations.

This Springer imprint is published by the registered company Springer Nature Switzerland AG
The registered company address is: Gewerbestrasse 11, 6330 Cham, Switzerland

If disposing of this product, please recycle the paper.

Preface

This book is the synthesis of lecture notes prepared for courses at the introductory graduate level jointly offered at the University of Padova and by one of us (R.O.) at Dartmouth. The main goal has been to introduce interested students to research topics in ultracold atomic physics, and more generally to expose scientists working in other areas of frontier physics to this novel and exciting research direction.

Ultracold atomic physics has been a major source of novel results since the first achievement of Bose-Einstein condensation in 1995. In spite of being around since almost three decades, a time lapse in which many subfields of physics become mature and even on the verge of morphing into other subfields, ultracold atom physics maintains its vitality and richness. This is due to various factors, from the large number of small-size experimental groups performing table-top experiments, to the various ramifications of the subfield in quantum and statistical physics, most notably many-body theory, quantum chemistry, and quantum metrology. One of the key features of the success of the subfield is also the relative easiness with which it is possible to join it. Knowledge of material usually mastered at the undergraduate level, such as electromagnetism, optics, nonrelativistic quantum mechanics, and statistical physics, is enough to at least understand the language and most of the literature of the subfield. It is in this context that we have decided to contribute with an agile textbook that could be understood by a senior undergraduate student or a first-year graduate student, regardless of their future specialization.

The vision we had is to provide material such that a student interested in ultracold atom physics could, after studying this book, start work in the laboratory, on theoretical aspects, or at least understand most of the titles of current literature and the more elaborated textbooks already available. In addition, we believe that this textbook could be used as a side reference in undergraduate courses on "conventional" condensed matter physics, as ultracold atom physics provides a platform in many ways more amenable to a fundamental understanding of condensed matter in the usual sense of the word. Due to its bottom-up approach, "synthetic matter" studied in ultracold atom physics allows to better enucleate specific model Hamiltonians otherwise difficult to study in the variegate solid or liquid-state environments provided to us by Nature, in which many competing effects can mask the desired dynamics.

The book is organized into five chapters. In the first, we recall fundamental aspects of equilibrium statistical mechanics, the statistical ensembles, and quantum statistics. The behavior of an ideal gas of Bose or Fermi type confined in a box and in a harmonic trap is discussed in detail. This is followed by the analysis of a weakly interacting and degenerate Bose gas in a mean-field approximation. In the second chapter we first discuss the window of opportunity to explore quantum degenerate gases under realistic constraints, and then discuss various cooling mechanisms, both based on single-particle interactions with photons—laser cooling—and interatomic interactions, such as evaporative and sympathetic cooling. In the third chapter, after describing imaging techniques, we discuss experiments admitting a quantitative interpretation in terms of mean-field theory approaches, first evidences for quantum degeneracy in Bose and Fermi gases, and more intriguing features, such as superfluidity and formation of topological defects. This allows us to also discuss analogies and differences with the case of superfluid bosonic Helium. In the fourth chapter we instead focus on strongly correlated systems, for which the description in terms of perturbative approaches fails. This includes general cases of pairing between fermions, fermion superfluidity and the crossover between the Bardeen-Cooper-Schrieffer (BCS) and the Bose-Einstein Condensate descriptions (BCS-BEC crossover), connections to superfluid fermionic Helium, and critical phenomena not contained within the Landau paradigm such as the Mott-insulator, the Berezinskii-Kosterlitz-Thouless phase transitions, as well as the physics of strongly interacting bosons under one-dimensional confinement, and transient critical phenomena. The fifth chapter is devoted to the implementation of ultracold atoms as a platform for quantum metrology and quantum computation and emulation. In general, these applications require non-interacting atoms, and the longest coherence times. Various appendices are then presented to complement the chapters or to contain explicit calculations too cumbersome to maintain in the main text. More specifically, two appendices are devoted to the second quantization (A) and its relationship of the Gross-Pitaevskii equation (B), one is centered on the theory of atomic scattering (C), and another to define quantities useful in a quantum information setting (D). At the end of each chapter we provide six exercises aimed to check the level of theoretical understanding and to acquire more familiarity with the orders of magnitude of the relevant physical quantities. Also included at the end of each chapter is a list of books and review papers. A necessarily partial and subjective list of original research papers, i.e., primary scholarship in the form of experiments, demonstrations, and theoretical ideas and techniques, is organized in the final bibliography. For practical reasons, the citations in the text are mentioned by the first author only, while in the bibliography all authors are quoted.

In line with the idea to swiftly introduce interested students to this subfield, we have deliberately adopted a more synthetic and logical approach in the presentation, delving into the historical sequence of events only when deemed strictly necessary, especially in connection to results already studied in condensed matter physics. As a consequence, we have omitted the detailed description of often tortuous paths of actual research, and therefore we feel not having carefully distributed proper credit to the relevant effort of many individuals, apologizing in advance for this choice and hoping in the understanding of our colleagues.

In terms of prerequisites, we assume some basic knowledge of quantum mechanics, statistical mechanics, electromagnetism, and optics at the level of the junior or senior years of the undergraduate studies. Quite involved discussions on results achieved more rigorously in a quantum mechanical setting are sometimes bypassed with their classical or semiclassical derivation, if existing. We have tried to avoid a "last minute" description of recent experiments for at least two reasons. The vastness of this subfield prevents any attempt in this direction within the boundaries we have self-imposed for the length of the book, and the fast pace of progress could make any description obsolete on a quite short timescale. Therefore we have been careful in not delving too much on the current frontier, spending instead significant effort to provide the reader with the tools required to understand future, often unpredictable, developments.

Any science book is the outcome of an uncountable number of opportunities of discussions with colleagues, and this monograph is no exception. First of all, we are grateful to all the students who have attended our course offers along about 20 years in two continents. With their questions and the need to provide homework and in-class discussion, we have progressively sharpened our views on these fascinating topics. We have also benefited from extensive discussions with many colleagues and collaborators, among these Rufus Boyack, Michael Brown-Hayes, Maria Luisa Chiofalo, Stephen Choi, Robin Cotè, Diego A. R. Dalvit, Jacek Dziarmaga, Vincent Flynn, Wolfgang Ketterle, Woo-Joong Kim, Giuseppe Carlo La Rocca, Francesco Lorenzi, Carlo Presilla, Chandra Raman, Hossein Sadeghpour, Bala Sundaram, Eddy Timmermans, Flavio Toigo, Lorenza Viola, Precooh Viola, Qun Wei, and Kevin C. Wright. Finally, we would like to acknowledge support from CNR (Consiglio Nazionale delle Ricerche), Dartmouth College, INFN (Istituto Nazionale di Fisica Nucleare), and the University of Padova.

Hanover, NH, USA
Padova, Italy
June 15, 2024

Roberto Onofrio
Luca Salasnich

Contents

1	**Quantum Degenerate Gases**		1
	1.1 Reminder of Equilibrium Statistical Mechanics		1
		1.1.1 Microcanonical Ensemble	4
		1.1.2 Canonical Ensemble	6
		1.1.3 Grand Canonical Ensemble	8
	1.2 Fermi-Dirac and Bose-Einstein Distributions		10
	1.3 Gas of Fermions		16
	1.4 Gas of Photons		20
	1.5 Gas of Massive Bosons and Bose-Einstein Condensation		22
		1.5.1 Noninteracting Bosons in a Harmonic Trap	26
		1.5.2 Noninteracting Fermions in a Harmonic Trap	28
	1.6 Interacting Condensates: Gross-Pitaevskii Equation		29
		1.6.1 Thomas-Fermi Approximation	32
		1.6.2 Gaussian Variational Approach	34
	1.7 Problems		35
	1.8 Further Reading		36
	References		37
2	**Trapping and Cooling of Atoms**		39
	2.1 The Temperature-Density Window for Quantum Degeneracy		39
	2.2 Reminder of Atomic Physics		41
	2.3 Doppler Cooling		46
	2.4 Magneto-Optical Trap		49
	2.5 Magnetic Traps		52
	2.6 Optical Dipole Traps		57
	2.7 Optical Lattices		61
	2.8 Evaporative and Sympathetic Cooling		63
	2.9 Problems		68
	2.10 Further Reading		69
	References		70
3	**Ultracold Atoms as Weakly Correlated Systems**		73
	3.1 Imaging of Ultracold Atomic Clouds		73
	3.2 First-Generation Experiments		76

	3.3	Quantum Transport Properties of Weakly Interacting Bose Gases	84
		3.3.1 Translational Response of a Bose-Einstein Condensate	87
		3.3.2 Rotational Response of a Bose-Einstein Condensate	88
		3.3.3 Experiments on Critical Velocities and Quantized Vortices	89
	3.4	Superfluidity in ^{4}He	91
	3.5	Topological Defects in Lower Effective Dimensionality	96
	3.6	Problems	101
	3.7	Further Reading	101
	References		102
4	**Ultracold Atoms as Strongly Correlated Systems**		**105**
	4.1	Interacting Fermi Systems	105
	4.2	BCS Model	109
		4.2.1 BCS Pairing at Finite Temperature	117
		4.2.2 Phase Diagram and Quantized Vortices	123
	4.3	Unconventional Phase Transitions	126
		4.3.1 Mott-Superfluid Quantum Phase Transition	127
		4.3.2 Berezinskii-Kosterlitz-Thouless Phase Transition	129
		4.3.3 Tonks-Girardeau Gas	133
		4.3.4 Dynamical Critical Phenomena: Kibble-Zurek Mechanism	134
	4.4	Problems	135
	4.5	Further Reading	135
	References		136
5	**Ultracold Atoms as Coherent Systems**		**139**
	5.1	Coherence of Photons and Matter Waves	140
	5.2	Atomic Interferometers	144
	5.3	Atom Lasers	151
	5.4	Superradiance and Matter-Wave Amplification	154
	5.5	Josephson-like Interferometry	157
	5.6	Atomtronics and Quantum Emulators	163
	5.7	Problems	167
	5.8	Further Reading	168
	References		168
Epilogue			**171**
A	**Second Quantization and Many-Body Systems**		**173**
B	**Coherent States and Gross-Pitaevskii Equation**		**183**
C	**Scattering Theory**		**187**
D	**Qubits and Entanglement**		**193**

Quantum Degenerate Gases

In this chapter, we discuss the main results of quantum statistical mechanics required for an understanding of the behavior of ultracold atomic gases, also assuming that the reader is already familiar with general principles of thermodynamics and statistical mechanics as from standard textbooks on these subjects. We first review basic concepts of equilibrium statistical mechanics with particular emphasis on the statistical ensembles and then proceed to study Bose and Fermi systems in terms of static and idealized properties, in the presence of spatial confinement. Finally, we discuss the dynamics of bosons in a single occupied state and in the presence of interatomic interactions using effective many-body time-dependent equations.

1.1 Reminder of Equilibrium Statistical Mechanics

Statistical mechanics aims at describing macroscopic properties of complex systems starting from their microscopic components. In doing this, it shares the same subjects of study as thermodynamics but differs because the latter assumes an effective description disregarding the atomic nature of matter, only focusing on macroscopic observables which, from the statistical mechanics standpoint, are representative of a coarse-grained dynamics. In principle, if a many-body quantum mechanical wave function is initially assigned, it is possible to solve the many-body Hamiltonian for the system to determine its time evolution and to access every possible observable at any time. Unfortunately, as in the classical counterpart, this is practically impossible for a large number of atoms, therefore a statistical approach is in order.

The maximal information on the state of a quantum system is obtained by using a vector $|\psi\rangle$, in the Dirac notation, of a Hilbert space $\mathcal{H}$. In a more realistic and general situation, the exact state of the system in the Hilbert space is unknown. This is a situation already known in classical mechanics. The maximal knowledge of a system in classical mechanics is achieved by specifying the location of the system in phase space, with well-defined positions and momenta. Due to possible impracticality, in a broad variety of contexts including many-body systems, to collect all this information, a coarse-grained description is required, by specifying the probability density in phase space. Even in the Hilbert space we must limit ourselves, in a similar context of partial knowledge of the system, to evaluate the probabilities for the occurrence of each state. When the probability distribution is nonzero on more than one state, the system is in a so-called mixed state, and a generalization of the usual description in terms of pure states $|\psi\rangle$ is then necessary, as we discuss next.

The average of a generic observable $\hat{A}$ on a state $|\psi\rangle$ is expressed as

$$\langle \hat{A} \rangle_\psi = \langle \psi | \hat{A} | \psi \rangle. \tag{1.1}$$

Let us consider an orthonormal basis in the Hilbert space $\{|\phi_i\rangle\}$ such that the generic state is written as $|\psi\rangle = \sum_i c_i |\phi_i\rangle$, with $c_i = \langle \phi_i | \psi \rangle$. Then, Eq. (1.1) becomes

$$\langle \hat{A} \rangle_\psi = \sum_{i,j} c_i c_j^* \langle \phi_j | \hat{A} | \phi_i \rangle = \sum_{i,j} \langle \phi_i | \psi \rangle \langle \psi | \phi_j \rangle \langle \phi_j | \hat{A} | \phi_i \rangle = \sum_{i,j} \rho_{ij} A_{ji}, \tag{1.2}$$

where $A_{ji} = \langle \phi_j | \hat{A} | \phi_i \rangle$, and we have introduced a new operator, called the density operator (or density matrix if specified in a concrete representation), whose matrix elements in the chosen basis ρ_{ij} are $\rho_{ij} = \langle \phi_i | \psi \rangle \langle \psi | \phi_j \rangle$. This operator then is

$$\hat{\rho} = |\psi\rangle \langle \psi|, \tag{1.3}$$

and the average of any observable is expressed, based on the last expression in Eq. (1.2), in terms of the density operator as

$$\langle \hat{A} \rangle_\rho = \text{Tr}[\hat{\rho} \hat{A}]. \tag{1.4}$$

Evidently, the density operator is normalized to unity, as manifest by using the identity operator $\hat{A} = \mathbb{1}$ in Eq. (1.4), $\text{Tr}[\hat{\rho}] = 1$. Another relevant property of the density operator is that it is a projector operator, $\hat{\rho}^2 = \hat{\rho}$, and then $\text{Tr}[\hat{\rho}^2] = 1$.

The density operator allows for a description of situations in which the state vector is known only probabilistically among a set of vectors $|\psi_\alpha\rangle$, each weighted by a probability p_α

$$\hat{\rho} = \sum_\alpha p_\alpha |\psi_\alpha\rangle \langle \psi_\alpha|, \tag{1.5}$$

where $\sum_\alpha p_\alpha = 1$. The density operator represents the most general description of a system, encoding two independent sources of uncertainty and related probabilities, the statistical one due to some coarse graining on potentially available information and the quantum one due to the intrinsic limitations imposed by the quantum theory

1.1 Reminder of Equilibrium Statistical Mechanics

in the knowledge of all possible observables. If we can avoid coarse graining and we know the unique state $|\psi\rangle$ accessible to the system, the density operator becomes simply $\hat{\rho} = |\psi\rangle\langle\psi|$. If the system can be considered classical, the density operator will become instead the corresponding probability density in phase space. The time evolution of the density operator can be derived by using Eq. (1.5) and the Schrödinger equation (with $\hbar$ the reduced Planck constant) for pure states, leading to

$$i\hbar \frac{d\hat{\rho}}{dt} = [\hat{H}, \hat{\rho}], \qquad (1.6)$$

which confirms the central role of the Hamiltonian operator in determining the dynamics, and indicates that a situation is stationary if

$$[\hat{H}, \hat{\rho}] = 0. \qquad (1.7)$$

This is relevant in our context because the simplest cases of application of statistical mechanics occur when, although the microscopic entities of the system are evolving in time, average values of macroscopic observables do not depend on time, and then, we speak of statistical mechanics at equilibrium. The equilibrium state is more easily accessible, rather than averaging the observables on the state at each instant of time, by considering a large number of replicas, or ensembles, of the system and averaging over these replicas according to the probability for each configuration to be realized in the system. The determination of the (time-independent) averages of macroscopic observables may then be obtained by sampling over "statistical ensembles", bypassing the need to know the detailed time evolution of the microscopic dynamics. The choice of statistical ensembles is not unique and depends on the interplay between the system under study and the external environment and their relationship to the Hamiltonian operator which, for equilibrium statistical mechanics, must fulfil Eq. (1.7). More specifically, we will discuss three statistical ensembles, microcanonical, canonical, and grand-canonical. A proper choice of the parameters in the three ensemble makes possible that the average values of macroscopic observables in general are identical in some idealized situation, such as the thermodynamic limit (defined, for an homogeneous system, as the limit $N, V \to \infty$ while keeping N/V constant). Nevertheless, the three ensembles have different physical meaning and are preferentially applied in different contexs as we describe in the following.

It should be noticed from the outset that the samples of ultracold atoms we will discuss later on are not, rigorously speaking, at equilibrium. They are quickly formed and remain for some time in a metastable state, after which they disappear for various reasons, for instance, through formation of other aggregates via three-body recombination or simply because they escape from the trapping region. Our description in terms of equilibrium statistical mechanics has to be considered as a first approximation valid only within the typical timescales in which relevant experiments on ultracold atoms are performed. This perceived limitation of ultracold atoms actually lends itself to provide a platform to study non-equilibrium quantum statistical mechanics, a subfield currently under development.

1.1.1 Microcanonical Ensemble

We start the analysis by considering a many-body quantum system with a fixed number N of identical particles characterized by the Hamiltonian operator $\hat{H}$ such that

$$\hat{H}|E_i\rangle = E_i|E_i\rangle , \qquad (1.8)$$

where $|E_i\rangle$ is the ith eigenstate of $\hat{H}$ and E_i is the corresponding energy eigenvalue. In the microcanonical ensemble, the quantum many-body system confined in a volume V has a fixed number N of particles and a fixed energy E. Such a system is also called an isolated system, with no possibility to trade particles and energy with the environment. In this case, the Hamiltonian $\hat{H}$ admits the spectral decomposition:

$$\hat{H} = \sum_i E_i |E_i\rangle\langle E_i| . \qquad (1.9)$$

The statistical description of the system is given by counting the number of distinct microstates corresponding to different eigenstates of the Hamiltonian operator and compatible with the same measurement of total energy E_i. The multiplicity of microstates (also called microcanonical volume) as a function of the energy E will be then

$$W = \sum_i \delta_{E,E_i} , \qquad (1.10)$$

where $\delta_{a,b}$ is the Kronecker delta symbol. All states with energy $E \neq E_i$ have zero probability, and each microstate with total energy $E = E_i$ has equal probability to occur. The multiplicity of states W in general will depend on various variables, such as the energy itself (in general more energy corresponds to higher multiplicity, at least for systems with infinite number of energy eigenstates), the volume in space, and internal properties of the particles, for instance. Also, although in the microcanonical ensemble the number of particles is considered constant, we can formally consider the dependence of the multiplicity of states upon this quantity. Based on the probabilities for each microscate, we can introduce a microcanonical density operator as

$$\hat{\rho} = \frac{1}{W}\delta_{E,\hat{H}} , \qquad (1.11)$$

1.1 Reminder of Equilibrium Statistical Mechanics

with spectral decomposition

$$\hat{\rho} = \sum_i \delta_{E,E_i} |E_i\rangle\langle E_i| \,. \tag{1.12}$$

The ensemble average of an observable described by the self-adjoint operator $\hat{A}$ is defined by Eq. (1.4). The connection with equilibrium thermodynamics is given by the formula of Ludwig Boltzmann (1872) for the statistical entropy:

$$S = k_B \ln W \,, \tag{1.13}$$

with $k_B = 1.38 \cdot 10^{-23}$ J/K the Boltzmann constant. This quantity represents the same concept as the microcanonical volume, but it has an additive character, so the entropy of two systems with no mutual interaction will be obtained by summing the individual entropies. From the entropy as a function of the macroscopic variables E, V, and N, i.e., $S(E, V, N)$, temperature T, pressure P, and chemical potential μ are obtained as

$$\frac{1}{T} = \left(\frac{\partial S}{\partial E}\right)_{V,N}, \quad P = T\left(\frac{\partial S}{\partial V}\right)_{E,N}, \quad \mu = -T\left(\frac{\partial S}{\partial N}\right)_{E,V}, \tag{1.14}$$

such that

$$dS = \left(\frac{\partial S}{\partial E}\right)_{V,N} dE + \left(\frac{\partial S}{\partial V}\right)_{E,N} dV + \left(\frac{\partial S}{\partial N}\right)_{E,V} dN = \frac{1}{T}dE + \frac{P}{T}dV - \frac{\mu}{T}dN \,. \tag{1.15}$$

By inverting Eq. (1.15), the energy E may be considered a function of the entropy, volume, and number of particles, i.e., $E = E(S, V, N)$, usually named internal energy. Its exact differential is

$$dE = TdS - PdV + \mu dN \,, \tag{1.16}$$

and therefore

$$T = \left(\frac{\partial E}{\partial S}\right)_{V,N}, \quad P = -\left(\frac{\partial E}{\partial V}\right)_{S,N}, \quad \mu = \left(\frac{\partial E}{\partial N}\right)_{S,V}. \tag{1.17}$$

In this setting, T, P, and μ play the role of complementary, dependent, thermodynamic variables obtained via partial derivatives of the internal energy with respect to the three independent variables S, V, and N. The choice of independent variables is not unique and depends upon the specific control an experimentalist has over the system. Legendre transformations allow to introduce a set of thermodynamical potentials with independent variables other than S, V, or N. In principle, for three independent variables, there are eight (2^3) thermodynamical potentials, although not all of them are used in practice. The existence of thermodynamical potentials admitting exact differentials, such that the sum of heat and work exchanged in an arbitrary transformation depends only on the initial and final thermodynamical states, is the content of the so-called first principle of thermodynamics. The latter allow to generalize the law of energy conservation for mechanical systems to thermal and chemical phenomena.

1.1.2 Canonical Ensemble

In the canonical ensemble, the quantum system in a volume V has a fixed number N of particles and a fixed temperature T, while it can exchange energy with the external world, the so-called heat bath or reservoir. The assignment of probabilities to each state is no longer uniform for all energy eigenstates but depends on the dimensionless parameter $E/(k_B T)$. For positive defined temperatures, states corresponding to lower energy eigenvalues are more likely than states corresponding to higher energy eigenvalues. Negative temperatures are also possible if the opposite occurs, as it happens in the case of population inversion in lasers or in nuclear spin systems.

The thermodynamical potential that is more adequate to describe this situation, since energy is not conserved, is obtained as the Legendre transformation of the internal energy with respect to entropy, also called Helmholtz free energy:

$$F(T, V, N) = E(S, V, N) - TS. \qquad (1.18)$$

Notice the important difference from the internal energy. For an isolated system, the internal energy is minimized by putting all particles in the ground state, which usually has zero statistical entropy since the ground state is degenerate, therefore corresponding to only one microstate. If a system is instead in contact with a reservoir at constant temperature, minimization of the Helmoltz free energy usually occurs if some particles are in excited states. Indeed, in this way, the entropy term in Eq. (1.18) is different from zero and, depending on temperature, can lower the free energy more than the net increase in internal energy due to the presence of excited particles.

The density operator in the canonical ensemble is quite different from the one in the microcanonical ensemble, because all possible energies E now have nonzero probabilities:

$$\hat{\rho} = \frac{1}{Z_N} e^{-\beta \hat{H}}, \qquad (1.19)$$

1.1 Reminder of Equilibrium Statistical Mechanics

where $\beta = 1/(k_B T)$ is the inverse temperature, with the dimensions of the reciprocal of the energy. We have also introduced a normalization factor Z_N, called canonical volume:

$$Z_N = \sum_i e^{-\beta E_i}, \tag{1.20}$$

where the sum is still extended, as in the microcanonical case, to all the energy eigenstates corresponding to the same energy eigenvalue E_i, and furthermore, it is summed over all possible energies. In this way, $\hat{\rho}$ continues to satisfy $\mathrm{Tr}(\hat{\rho}) = 1$. The canonical volume is a function of various macroscopic quantities and for this reason is also called partition function. The density operator $\hat{\rho}$ has the spectral decomposition:

$$\hat{\rho} = \sum_i e^{-\beta E_i} |E_i\rangle\langle E_i|, \tag{1.21}$$

and the canonical partition function is expressed in terms of the density operator as

$$Z_N = \mathrm{Tr}[e^{-\beta \hat{H}}]. \tag{1.22}$$

The ensemble average of a generic observable $\hat{A}$ is defined as in Eq. (1.4). In the specific case of the internal energy, its average value is obtained as

$$E = \langle \hat{H} \rangle = \frac{1}{Z_N} \sum_i E_i e^{-\beta E_i} = -\frac{1}{\beta} \frac{\partial \ln(Z_N)}{\partial \beta}. \tag{1.23}$$

This relationship, together with the definition Eq. (1.18) of the Helmholtz free energy in terms of the internal energy, yields

$$Z_N = e^{-\beta F}. \tag{1.24}$$

The exact differential for the Helmholtz free energy $F(T, V, N)$ is

$$dF = \left(\frac{\partial F}{\partial T}\right)_{V,N} dT + \left(\frac{\partial F}{\partial V}\right)_{T,N} dV + \left(\frac{\partial F}{\partial N}\right)_{T,V} dN = -S dT - P dV + \mu dN \tag{1.25}$$

which allows to identify entropy S, pressure P, and chemical potential μ as

$$S = -\left(\frac{\partial F}{\partial T}\right)_{V,N}, \quad P = -\left(\frac{\partial F}{\partial V}\right)_{T,N}, \quad \mu = \left(\frac{\partial F}{\partial N}\right)_{T,V}. \tag{1.26}$$

Indeed, in the canonical ensemble, the independent thermodynamic variables are T, N, and V, while S, P, and μ are dependent thermodynamic variables. The formalism allows to evaluate energy fluctuations around the average energy of the system. In general, we expect these fluctuations to become negligible in the limit of a large number of particles, i.e., $\Delta E/E \to 0$ as $N \to \infty$, in the thermodynamic limit. In this situation, there is no difference from the analogous quantity evaluated in the microcanonical ensemble in which $\Delta E = 0$ by definition.

1.1.3 Grand Canonical Ensemble

In the grand canonical ensemble, neither the energy nor the number of particles of the system in a volume V are conserved. The system can trade energy and particles with the external world. This leads to the introduction of number operators: $\hat{N}_i$ for each energy level and the total number operator

$$\hat{N} = \sum_i \hat{N}_i = \sum_i n_i |N_i\rangle\langle N_i|, \tag{1.27}$$

with n_i and $|N_i\rangle$ the eigenvalue and eigenstate, respectively, of the number operator $\hat{N}_i$

$$\hat{N}_i |N_i\rangle = n_i |N_i\rangle, \tag{1.28}$$

and N the eigenvalue of the total number operator. However, unlike the canonical case, here, the total number of particles is not fixed.

This situation is important, in thermodynamics, to describe chemical reactions. The system, if at equilibrium, has a fixed temperature T and a fixed chemical potential μ.

For the grand canonical ensemble, one defines the grand canonical density operator as

$$\hat{\rho} = \frac{1}{\mathcal{Z}} e^{-\beta(\hat{H} - \mu \hat{N})}, \tag{1.29}$$

where the normalization factor $\mathcal{Z}$ is defined as

$$\mathcal{Z} = \sum_{N=0}^{\infty} \sum_i e^{-\beta(E_i - \mu N)}. \tag{1.30}$$

with a double sum over all possible energy eigenstates and all possible number of particles. The density operator $\hat{\rho}$ has the spectral decomposition:

$$\hat{\rho} = \sum_{N=0}^{\infty} \sum_i e^{-\beta(E_i - \mu N)} |E_i\rangle\langle E_i| = \sum_{N=0}^{\infty} z^N \sum_i e^{-\beta E_i} |E_i\rangle\langle E_i|, \tag{1.31}$$

1.1 Reminder of Equilibrium Statistical Mechanics

where $z = e^{\beta\mu}$ is named fugacity and the grand canonical partition function can be expressed in terms of the corresponding density operator as

$$\mathcal{Z} = \text{Tr}[e^{-\beta(\hat{H}-\mu\hat{N})}]. \tag{1.32}$$

The grand canonical potential is defined as

$$\Omega(T, V, \mu) = F(T, V, N) - \mu N, \tag{1.33}$$

related to the grand canonical partition function by the relatioship

$$\mathcal{Z} = e^{-\beta\Omega}. \tag{1.34}$$

The exact differential of $\Omega(T, V, \mu)$ is

$$d\Omega = \left(\frac{\partial\Omega}{\partial T}\right)_{V,\mu} dT + \left(\frac{\partial\Omega}{\partial V}\right)_{T,\mu} dV + \left(\frac{\partial\Omega}{\partial \mu}\right)_{T,V} d\mu = -SdT - PdV - Nd\mu \tag{1.35}$$

allowing to express entropy S, pressure P, and the average number N of particles as

$$S = -\left(\frac{\partial\Omega}{\partial T}\right)_{V,\mu}, \quad P = -\left(\frac{\partial\Omega}{\partial V}\right)_{T,\mu}, \quad N = -\left(\frac{\partial\Omega}{\partial \mu}\right)_{V,T}. \tag{1.36}$$

In the grand canonical ensemble, the independent thermodynamic variables are T, V, and μ, while S, P, and N are dependent thermodynamic variables.

To conclude this section, we observe that in the grand canonical ensemble, instead of working with eigenstates $|E_i\rangle$ of $\hat{H}$ at fixed number N of particles, one can work with Fock states (see Appendix A):

$$|n_0 n_1 n_2 \ldots\rangle = |n_0\rangle \otimes |n_1\rangle \otimes |n_2\rangle \otimes \ldots, \tag{1.37}$$

where $|n_\alpha\rangle$ is the single-mode Fock state which describes n_α particles in the single-mode state $|\alpha\rangle$ with $\alpha = 0, 1, 2, \ldots$. Vice versa, one can work within the framework of the canonical ensemble then summing each canonical partition function over all possible number of particles weighted by the fugacity, from Eq. (1.31), as

$$\mathcal{Z} = \sum_{N=0}^{\infty} z^N Z_N. \tag{1.38}$$

This also shows that the same considerations discussed for the equivalence of the results between the microcanonical and the canonical ensemble are also applicable here to the equivalence between the canonical and the grand canonical

Table 1.1 Summary of some relevant quantities for the three statistical ensembles discussed in the text. We outline the quantities required for the equilibrium state of one or more subsystems, the thermodynamical potential which is minimized under equilibrium, the quantity summed over all microstates, and the "fundamental" relationship between the latter and the relevant thermodynamical potential, from which all statistical quantities at equilibrium, including their fluctuations, can be evaluated

Statistical ensemble	Microcanonical	Canonical	Grand canonical
Equilibrium	Maximum S	Equal T	Equal T and μ
Relevant potential	$E(S, V, N)$	$F(T, V, N)$	$\Omega(T, V, \mu)$
Sum over microstates	W	Z_N	$\mathcal{Z}$
Fundamental relationship	$W = e^{S/k_B}$	$Z_N = e^{-F/(k_B T)}$	$\mathcal{Z} = e^{-\Omega/(k_B T)}$

ensemble, as long as we are concerned with energy fluctuations. On top of this, the thermodynamical limit also implies, in general, equivalence between the canonical and the grand canonical ensemble in regard to number fluctuations, with the notable exception of critical phenomena, in which case large number fluctuations can be essential, as observed in the phenomenon of critical opalescence.

Also, in the specific situation of a continuum of quantum states, one recovers classical statistical mechanics, with the additional feature of incorporating the Planck constant as a convenient unit of measure of the volume in phase space, whose infinitesimal and adimensional volume is therefore $d^3 q_i d^3 p_i / \hbar^3$ for a single degree of freedom. In the classical limit, the density operator morphs into the phase space probability density $\rho(q_i, p_i)$ for a dynamical system characterized by the coordinates (q_i, p_i) in phase space. In this situation, the dynamics is ruled by the analogous of Eq. (1.6) for a classical system:

$$\frac{d\rho}{dt} = \frac{\partial \rho}{\partial t} + \{H, \rho\}. \tag{1.39}$$

The choice of a phase space probability density not explicitly depending on time and depending on (q_i, p_i) only through the Hamiltonian $H(q_i, p_i)$ is a sufficient condition to ensure its time independence; therefore, the considerations made for the three statistical ensembles discussed above hold also in the classical limit.

A synthesis of relevant quantities and relationships for the three statistical ensembles is presented in Table 1.1.

1.2 Fermi-Dirac and Bose-Einstein Distributions

Quantum mechanics affects the collective behavior of a many-body system even in the absence of interparticle interactions, due to the requirement of indistinguishability of identical particles. Basically, this stems from the uncertainty principle, since it is impossible to continuously monitor the trajectory of particles without perturbing their motion. In the presence of identical particles, this creates ambiguity on tracking

1.2 Fermi-Dirac and Bose-Einstein Distributions

down each of them, and therefore, the many-body wave function should satisfy symmetry properties under the exchange of any pair (or any subset) of particles. For a quantitative analysis, we introduce the generalized coordinate $x = (\mathbf{r}, \sigma)$ of a single particle taking into account the spatial coordinate $\mathbf{r}$ and the intrinsic spin σ pertaining to the particle. For instance, a spin 1/2 particle has spin eigenvalues along a given component as $\sigma = -1/2, 1/2$, in units of $\hbar$. By using the Dirac notation, the corresponding single-particle state is

$$|x\rangle = |\mathbf{r}\,\sigma\rangle. \tag{1.40}$$

We now consider N identical particles, i.e., particles with the same mass, electric charge, and any possible quantum number associate with their state. The many-body wave function of the system is given by

$$\Psi(x_1, x_2, \ldots, x_N) = \Psi(\mathbf{r}_1, \sigma_1, \mathbf{r}_2, \sigma_2, \ldots, \mathbf{r}_N, \sigma_N), \tag{1.41}$$

The requirement of indistinguishability translates into the invariance of the square modulus of this wave function under exchange of any two particles:

$$|\Psi(x_1, x_2, \ldots, x_i, \ldots, x_j, \ldots, x_N)|^2 = |\Psi(x_1, x_2, \ldots, x_j, \ldots, x_i, \ldots, x_N)|^2, \tag{1.42}$$

stating that probability of finding the particles must be independent on the exchange of two generalized coordinates x_i and x_j.

Experiments suggest that there are only two kinds of identical particles which satisfy Eq. (1.42): bosons and fermions. For N identical bosons, one has

$$\Psi(x_1, x_2, \ldots, x_i, \ldots, x_j, \ldots, x_N) = \Psi(x_1, x_2, \ldots, x_j, \ldots, x_i, \ldots, x_N), \tag{1.43}$$

i.e., the many-body wave function is symmetric with respect to the exchange of two coordinates x_i and x_j. For the simplest case of two identical bosons, this implies

$$\Psi(x_1, x_2) = \Psi(x_2, x_1). \tag{1.44}$$

For N identical fermions, one has instead

$$\Psi(x_1, x_2, \ldots, x_i, \ldots, x_j, \ldots, x_N) = -\Psi(x_1, x_2, \ldots, x_j, \ldots, x_i, \ldots, x_N), \tag{1.45}$$

i.e., the many-body wave function is antisymmetric with respect to the exchange of two coordinates x_i and x_j. For two identical fermions, this implies

$$\Psi(x_1, x_2) = -\Psi(x_2, x_1). \tag{1.46}$$

An immediate consequence of the antisymmetry of the fermionic many-body wave function is that if $x_i = x_j$ the many-body wave function is zero. In other words, the probability of finding two fermions with the same generalized coordinates is zero. Historically, this consequence of the antisymmetry for fermions was first inferred by Wolfgang Pauli to interpret helium spectroscopy and then assumed as a principle (Pauli principle) in the framework of the old quantum theory [16, 17]. In the "new" quantum mechanics, this is not a principle, rather just a simple corollary of the antisymmetric character of the many-body wave function for identical fermions. Nevertheless, the tradition is still there to call it "Pauli principle."

Also, a remarkable consequence of the extension of quantum mechanics in the relativistic realm, quantum field theory, is the *spin-statistics theorem*: Identical particles with integer spin are bosons, while identical particles with semi-integer spin are fermions. For instance, photons are bosons with spin 1, while electrons are fermions with spin 1/2. Notice that for a composite particle the statistics is determined by the total spin. For example, the total spin (sum of nuclear and electronic spins) of ^{4}He atom in its ground state is 0 and consequently is a boson, while the total spin 1/2 of ^{3}He qualifies this isotope as a fermion.

The Hamiltonian operator of N identical noninteracting particles of mass m is given by

$$\hat{H} = \sum_{i=1}^{N} \hat{h}(x_i) , \qquad (1.47)$$

where $\hat{h}(x)$ is the Hamiltonian operator for the ith particle. In the following, we work in a nonrelativistic setting and assume

$$\hat{h}(x_i) = -\frac{\hbar^2}{2m} \nabla_i^2 + U(x_i) , \qquad (1.48)$$

with U the potential energy due to an external field. The single-particle Hamiltonian operator $\hat{h}$ satisfies the eigenvalue equation:

$$\hat{h}(x) \phi_\alpha(x) = \epsilon_\alpha \phi_\alpha(x) , \qquad (1.49)$$

where ϵ_α are the single-particle energy eigenstates and $\phi_\alpha(x)$ are the single-particle eigenfunctions, with $\alpha = 0, 1, 2, 3, \ldots$ progressively denoting higher energy eigenvalues ($\epsilon_0 < \epsilon_1 \leq \epsilon_2 \ldots$). The many-body wave function $\Psi(x_1, x_2, \ldots, x_N)$ of the system can be written in terms of the single-particle wave functions $\phi_\alpha(x)$, but one must take into account the spin-statistics of the identical particles. Let us discuss the important case of the ground state, which is the many-body state with the minimum allowed total energy.

The many-body wave function enforcing the symmetry under particle exchange describing identical bosons in the ground state reads

$$\Psi(x_1, x_2, \ldots, x_N) = \phi_0(x_1) \phi_0(x_2) \ldots \phi_0(x_N) , \qquad (1.50)$$

1.2 Fermi-Dirac and Bose-Einstein Distributions

which corresponds to the configuration where all the particles are in the lowest-energy single-particle state $\phi_0(x)$. We stress that for noninteracting identical particles the Hamiltonian (1.47) is separable and the total energy associated to the bosonic many-body wave function (1.50) is

$$E = N \epsilon_0 . \tag{1.51}$$

For fermions, the many-body wave function describing the ground state is instead very different:

$$\Psi(x_1, x_2, \ldots, x_N) = \frac{1}{\sqrt{N!}} \begin{vmatrix} \phi_0(x_1) & \phi_0(x_2) & \cdots & \phi_0(x_N) \\ \phi_1(x_1) & \phi_1(x_2) & \cdots & \phi_1(x_N) \\ \cdots & \cdots & \cdots & \cdots \\ \phi_{N-1}(x_1) & \phi_{N-1}(x_2) & \cdots & \phi_{N-1}(x_N) \end{vmatrix} , \tag{1.52}$$

the so-called Slater determinant of the $N \times N$ matrix obtained with the N lowest-energy single-particle wave functions $\psi_\alpha(x)$, with $\alpha = 0, 1, 2, \ldots, N-1$, evaluated for the N generalized coordinates x_i, with $i = 1, 2, \ldots, N$. Note that for two fermions the previous expression becomes

$$\Psi(x_1, x_2) = \frac{1}{\sqrt{2}} [\phi_0(x_1)\phi_1(x_2) - \phi_0(x_2)\phi_1(x_1)] . \tag{1.53}$$

For the fermionic many-body wave function (1.52), the total energy (in the absence of degenerate single-particle energy levels) reads

$$E = \epsilon_0 + \epsilon_1 + \ldots + \epsilon_{N-1} , \tag{1.54}$$

which is surely higher than the bosonic one. The highest occupied single-particle energy level is called Fermi energy, and it is indicated as $\epsilon_F = \epsilon_{N-1}$.

The comparison between the ground states of a system of noninteracting bosons and fermions is depicted in Fig. 1.1 in the representative example of their

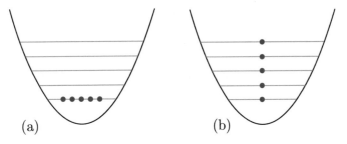

Fig. 1.1 Ground state of a system of identical noninteracting bosons (**a**) and fermions (**b**) in a harmonic trap

confinement in a harmonic trap. In both cases, the entropy is zero, as expected for a ground state. In the case of bosons, this is due to the fact that all particles are in the minimum energy state. Notice that one would have the same outcome also for a system of (classical) distinguishable particles if the ground state is nondegenerate. For fermions, the indistinguishability makes sure that, although the particles cannot share the same energy eigenstates, there is only one realization of the many-body ground state, unlike the case of classical particles in the same configuration.

Let us now consider the system in thermal and chemical equilibrium with a bath at a finite temperature T, described by a Hamiltonian operator $\hat{N}$ and a total number operator $\hat{N} = \sum_\alpha \hat{N}_\alpha$. The relevant quantity to calculate all thermodynamical properties of the system is the grand canonical partition function $\mathcal{Z}$, as defined in Eq. (1.30).

By performing the trace with the Fock states introduced in Eq. (1.37)

$$\mathcal{Z} = \sum_{\{n_\alpha\}} \langle \ldots n_\alpha \ldots | e^{-\beta(\hat{H}-\mu\hat{N})} | \ldots n_\alpha \ldots \rangle = \sum_{\{n_\alpha\}} e^{-\beta \sum_\alpha (\epsilon_\alpha - \mu) n_\alpha}$$

$$= \prod_\alpha \sum_{n_\alpha} e^{-\beta(\epsilon_\alpha - \mu) n_\alpha} . \tag{1.55}$$

In the case of bosons, any arbitrary number of particles can be in any energy eigenstate, resulting in a series which can be resummed if $\mu < \epsilon_\alpha$ which is satisfied for sure if $\mu < \epsilon_0$. Under this assumption, we get

$$\mathcal{Z}_- = \prod_\alpha \sum_{n=0}^\infty e^{-\beta(\epsilon_\alpha - \mu) n} = \prod_\alpha \frac{1}{1 - e^{-\beta(\epsilon_\alpha - \mu)}} . \tag{1.56}$$

Instead, for fermions, the sum is restricted to zero or one-particle occupancy:

$$\mathcal{Z}_+ = \prod_\alpha \sum_{n=0}^1 e^{-\beta(\epsilon_\alpha - \mu) n} = \prod_\alpha \left(1 + e^{-\beta(\epsilon_\alpha - \mu)}\right) . \tag{1.57}$$

This allows to evaluate the grand canonical potential:

$$\Omega_\mp = -\frac{1}{\beta} \ln(\mathcal{Z}_\mp) = -\frac{1}{\beta} \sum_\alpha \ln\left(1 \mp e^{-\beta(\epsilon_\alpha - \mu)}\right) \tag{1.58}$$

where $-$ is for bosons and $+$ for fermions. From the third relationship in Eq. (1.36), finally, we obtain

$$N_\alpha = \frac{1}{e^{\beta(\epsilon_\alpha - \mu)} \mp 1} . \tag{1.59}$$

expressing the Bose-Einstein distribution [1] and the Fermi-Dirac distribution [4,8].

1.2 Fermi-Dirac and Bose-Einstein Distributions

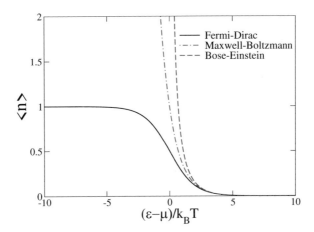

Fig. 1.2 Average number of particles $\langle n \rangle$ per state versus the scaled energy $(\epsilon - \mu)/(k_B T)$ for the three cases of bosons, fermions, and distinguishable particles, in the continuum approximations of states and at nonzero temperature, with $k_B T/\mu = 10^{-1}$. Notice that in the case of the Fermi-Dirac statistics the average number of particles is approaching 1 at low energies and 0 at high energies, while it is intermediate for energies differing from the chemical potential by few units of $k_B T$. For instance, if $\epsilon = 0.8\,\mu$ (i.e., it differs from μ by $-2k_B T$), we have $\langle n \rangle (0.8\,\mu) \simeq 0.88$, and if $\epsilon = 1.2\,\mu$ (i.e., *it* differs from μ by $+2k_B T$), we have $\langle n \rangle (1.2\,\mu) \simeq 0.12$ (in general the Fermi-Dirac distribution has a symmetry around the chemical potential, $\langle n(\epsilon - \mu) \rangle = 1 - \langle n(\epsilon + \mu) \rangle$). Therefore, the dynamics of fermions at finite temperature is frozen for all energies apart from the ones differing from the chemical potential by few units of $k_B T$

For a more practical discussion, we consider a limit in which the energy eigenvalues can be approximated in the continuum. This is a sort of semiclassical approximation, and it is justified in the limit of temperatures large enough with respect to the energy spacing between consecutive levels, a condition satisfied for all ultracold gases so far investigated. In this approximation, we replace $\epsilon_\alpha \to \epsilon$ and, as visibile in Fig. 1.2, the average number $\langle n \rangle$ of particles per state crucially depends on the statistics. In the case of fermions, $\langle n \rangle$ cannot be larger than one, and in the limit $T \to 0$, we get $\langle n \rangle = \Theta(\mu - \epsilon)$ with $\Theta(x)$ the Heaviside step function; each state is filled with one particle up to exhaustion of the particles. In the case of bosons, $\langle n \rangle$ diverges as $\mu \to \epsilon$; therefore, bosons tend to pile up in the ground state with unlimited average number. In the high energy limit, the exponential dependence on ϵ in Eq. (1.59) dominates with respect to unity, and both Bose-Einstein and Fermi-Dirac distributions tend to the classical Maxwell-Boltzmann distribution $\langle n \rangle = e^{-\beta(\epsilon - \mu)}$, still maintaining the average population of boson higher than the one the fermions at the same temperature. As discussed in the caption of Fig. 1.2, most of the fermions are frozen in their state, only the ones close to the chemical potential withing few units of $k_B T$ can make transitions to different energy levels.

This behavior has to be complemented by the evaluation of how many states are available for each energy level, also named density of states. This depends on the specific system, the dispersion relationship of its particles, and its effective dimensionality, and we will discuss in the following the two relevant examples of particles confined in a box and in a harmonic oscillator potential.

1.3 Gas of Fermions

A prototypical physical system is a noninteracting Fermi gas confined in a cubic volume V of side L by means of a three-dimensional infinite square well. The atoms do not experience any confining force until they hit one of the walls of the hollow cube. This case is simple to treat analytically and corresponds to a first rough model of free electrons in a conductor of cubic shape. It has also been demonstrated by confining ultracold atoms in a homogeneous potential [11].

The single-particle energy eigenstates inside the cube are three-dimensional plane waves:

$$\phi(x) = \frac{1}{\sqrt{V}} e^{i\mathbf{k}\cdot\mathbf{r}} \chi_\sigma , \qquad (1.60)$$

where χ_σ is the spin state along a chosen quantization axis, usually considered the "third" component of the spin. Outside the cube, under the hypothesis of ideal confinement inside, the energy eigenstates are made of a constant function zero everywhere. We now focus on the simplest case of spin 1/2 fermions, with eigenstates of the total spin operator and its third component written as

$$\chi_\uparrow = \begin{pmatrix} 1 \\ 0 \end{pmatrix}, \qquad \chi_\downarrow = \begin{pmatrix} 0 \\ 1 \end{pmatrix}. \qquad (1.61)$$

We impose continuity conditions for the energy eigenstates on all the walls of the cube, obtaining

$$e^{ik_x L} = 1, \quad e^{ik_y L} = 1, \quad e^{ik_z L} = 1. \qquad (1.62)$$

It follows that the linear momentum $\mathbf{k}$ can only take on the values:

$$k_x = \frac{2\pi}{L} n_x, \quad k_y = \frac{2\pi}{L} n_y, \quad k_z = \frac{2\pi}{L} n_z, \qquad (1.63)$$

where n_x, n_y, n_z are integer quantum numbers different from zero. The single-particle energies are given by

$$\epsilon_\mathbf{k} = \frac{\hbar^2 k^2}{2m} = \frac{\hbar^2}{2m} \frac{4\pi^2}{L^2} \left(n_x^2 + n_y^2 + n_z^2 \right), \qquad (1.64)$$

with $k^2 = k_x^2 + k_y^2 + k_z^2$.

1.3 Gas of Fermions

Working at zero temperature, the total number N and the total energy E of the particles are given by

$$N = \sum_\sigma \sum_{\mathbf{k}} \Theta(\epsilon_F - \epsilon_{\mathbf{k}}), \tag{1.65}$$

$$E = \sum_\sigma \sum_{\mathbf{k}} \epsilon_{\mathbf{k}} \Theta(\epsilon_F - \epsilon_{\mathbf{k}}), \tag{1.66}$$

where we also include the sum over the possible spin states. In the large size limit $L \to \infty$, the allowed values are closely spaced, and one can use the continuum approximation:

$$\sum_{n_x, n_y, n_z} \to \int dn_x\, dn_y\, dn_z, \tag{1.67}$$

which implies

$$\sum_{\mathbf{k}} \to \frac{L^3}{(2\pi)^3} \int d^3\mathbf{k} = V \int \frac{d^3\mathbf{k}}{(2\pi)^3}. \tag{1.68}$$

Therefore, we have

$$N = \sum_{\sigma=\uparrow,\downarrow} V \int \frac{d^3\mathbf{k}}{(2\pi)^3} \Theta\left(\epsilon_F - \frac{\hbar^2 k^2}{2m}\right), \tag{1.69}$$

from which one gets (with the spin degeneracy yielding a further factor 2) the uniform density

$$n = \frac{N}{V} = \frac{1}{3\pi^2} \left(\frac{2m\epsilon_F}{\hbar^2}\right)^{3/2}. \tag{1.70}$$

The formula allows to express the Fermi energy ϵ_F as a function of the density n, namely,

$$\epsilon_F = \frac{\hbar^2}{2m} \left(3\pi^2 n\right)^{2/3} = \frac{\hbar^2 k_F^2}{2m}, \tag{1.71}$$

where $k_F = \left(3\pi^2 n\right)^{1/3}$ is the so-called Fermi wave number.

For the energy, we have instead

$$E = \sum_{\sigma=\uparrow,\downarrow} V \int \frac{d^3\mathbf{k}}{(2\pi)^3} \frac{\hbar^2 k^2}{2m} \Theta\left(\epsilon_F - \frac{\hbar^2 k^2}{2m}\right), \tag{1.72}$$

from which the energy density is written

$$\mathcal{E} = \frac{E}{V} = \frac{3}{5} n \epsilon_F = \frac{3}{5} \frac{\hbar^2}{2m} \left(3\pi^2\right)^{2/3} n^{5/3} , \tag{1.73}$$

in terms of the Fermi energy ϵ_F and the density n.

The Fermi energy increases with the density of the system, which implies that a Fermi system is quantum degenerate, and nearly ideal, as long as this Fermi energy overwhelms any other energy scale, including the average thermal energy of order $k_B T$, i.e., $\epsilon_F \gg k_B T$. Indeed, one can introduce the Fermi temperature T_F as $T_F = \epsilon_F / k_B$, and the Fermi degeneracy holds for $T \ll T_F$. As expected, particles with smaller mass have larger Fermi energies, consistently with the general idea that quantum mechanics is more relevant, with other quantities being the same, for less massive particles. For instance, a metal contains a nearly Fermi degenerate gas of electrons already at room temperature, as the Fermi temperature is $T_F \simeq 10^4$ K.

Taking explicitly into account the temperature, the previous formulas for the total number N and the total energy E of the particles must be extended. Their finite-temperature version is given by

$$N = \sum_\sigma \sum_\mathbf{k} \frac{1}{e^{\beta(\epsilon_\mathbf{k} - \mu)} + 1} , \tag{1.74}$$

$$E = \sum_\sigma \sum_\mathbf{k} \epsilon_\mathbf{k} \frac{1}{e^{\beta(\epsilon_\mathbf{k} - \mu)} + 1} , \tag{1.75}$$

where, with respect to the $T = 0$ case, the Fermi-Dirac distribution substitutes the Heaviside step function and the chemical potential μ replaces the Fermi energy ϵ_F. After integration over wave vectors in the continuum, Eq. (1.68), one finds, assuming spin degeneracy with multiplicity g_s

$$n = \frac{g_s}{2\pi^3} \int d^3\mathbf{k} \frac{1}{e^{\beta(\epsilon_\mathbf{k} - \mu)} + 1} = \frac{g_s}{\lambda_T^3} f_{3/2}(z) \tag{1.76}$$

$$\mathcal{E} = \frac{g_s}{2\pi^3} \int d^3\mathbf{k} \frac{\epsilon_\mathbf{k}}{e^{\beta(\epsilon_\mathbf{k} - \mu)} + 1} = \frac{3}{2} g_s \frac{k_B T}{\lambda_T^3} f_{5/2}(z) \tag{1.77}$$

with $\epsilon(\mathbf{k})$ as in Eq. (1.64), z the fugacity, λ_T the de Broglie wavelength at temperature T

$$\lambda_T = \left(\frac{2\pi \hbar^2}{m k_B T}\right)^{1/2} , \tag{1.78}$$

1.3 Gas of Fermions

and $f_n(z)$ the Fermi function, defined as

$$f_n(z) = z \frac{\int_0^{+\infty} dx \, \frac{x^{n-1}}{z^{-1}e^x + 1}}{\int_0^{+\infty} dx \, \frac{x^{n-1}}{z^{-1}e^x}} = \sum_{k=1}^{+\infty} (-1)^{k+1} \frac{z^k}{k^n}. \qquad (1.79)$$

Notice that in Eq. (1.79) the quantity in the denominator of the intermediate relation is the gamma function $\Gamma(n)$ multiplied by z. The numerator tends to a gamma function when $z \ll 1$, the high-temperature limit, therefore in the same limit $f_n(z)$ tends to z. This is also confirmed by considering the first terms of the series in the last expression of Eq. (1.79).

Equation (1.76) enables to determine numerically the chemical potential μ as a function of the temperature T at fixed number density n. However, in the low-temperature regime, where $k_B T \ll \epsilon_F$, from Eq. (1.76) one can Taylor expand the expression to find the next-to-leading term to the zero-temperature result $\mu = \epsilon_F$ (Sommerfeld expansion), namely,

$$\mu \simeq \epsilon_F - \frac{\pi^2}{12} \frac{(k_B T)^2}{\epsilon_F}. \qquad (1.80)$$

Similarly, from Eqs. (1.76) and (1.77), one also finds

$$\mathcal{E} \simeq \frac{3}{5} n \epsilon_F - \frac{\pi^2}{4} \frac{n(k_B T)^2}{\epsilon_F}. \qquad (1.81)$$

In the experiments with ultracold atomic gases, a relevant quantity that is often measured at constant volume V and fixed number N of particles is the specific heat, defined as

$$c_V = \frac{1}{N} \left(\frac{\partial E}{\partial T} \right)_{V,N}. \qquad (1.82)$$

In Fig. 1.3, we plot the specific heat c_V as a function of the temperature. At low temperatures, it goes to zero, as required by the third principle of thermodynamics assigning zero entropy for a nondegenerate ground state. At high temperatures, $c_V/(Nk_B)$ approaches $3/2$, consistent with the expectation from classical statistical mechanics that the energy is uqually shared among all available degrees of freedom, the equipartition theorem. In the intermediate regime of temperatures, the number of fermions available for hopping to other energy levels, as commented in the caption of Fig. 1.2, is proportional to $k_B T$, and they can trade $\simeq k_B T$ at most; therefore, the energy exchanged is proportional to T^2 implying, based on Eq. (1.82), a linear dependence of the specific heat on temperature.

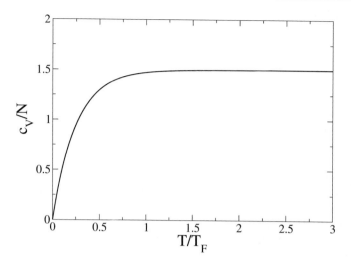

Fig. 1.3 Specific heat per atom in units of the Boltzmann constant k_B, for an ideal Fermi gas confined in a three-dimensional box, versus the temperature T normalized to the Fermi temperature T_F

1.4 Gas of Photons

Let us consider a gas of photons, which are massless bosons with spin 1, in thermal equilibrium with the external environment, for instance, in a cavity or an oven, at the temperature T [1]. As previously stressed, the relevant quantity to calculate all thermodynamical properties of the system is the grand canonical partition function of Eq. (1.30) using the number operator and the Hamiltonian operator:

$$\hat{N} = \sum_s \sum_{\mathbf{k}} \hat{N}_{\mathbf{k}s}, \tag{1.83}$$

$$\hat{H} = \sum_s \sum_{\mathbf{k}} \hbar \omega_k \hat{N}_{\mathbf{k}s}, \tag{1.84}$$

where $\mathbf{k}$ is the wavenumber, s denotes the polarization states, two for massless spin 1 particles like photons, and we have neglected the temperature-independent zero-point photons and their associated energy, corresponding in quantum field theory to implement the formal procedure of normal ordering. Another consequence of the massless nature of photons is that the minimal energy required to create a photon may be arbitrarily small, and consequently, the chemical potential μ is zero. In other words, apart from special counterexamples, the number of photons is not fixed, and

1.4 Gas of Photons

photons can be created and destroyed without paying any price. Then, the grand canonical partition function Eq. (1.56) becomes

$$\mathcal{Z} = \prod_{s\mathbf{k}} \frac{1}{1 - e^{-\beta \hbar \omega_k}} \,. \tag{1.85}$$

This allows to calculate the average number of photons and their average energy:

$$N = \sum_s \sum_{\mathbf{k}} \langle \hat{N}_{\mathbf{k}s} \rangle = \sum_s \sum_{\mathbf{k}} \frac{1}{e^{\beta \hbar \omega_k} - 1}, \tag{1.86}$$

$$E = \sum_s \sum_{\mathbf{k}} \hbar \omega_{\mathbf{k}} \langle \hat{N}_{\mathbf{k}s} \rangle = \sum_s \sum_{\mathbf{k}} \frac{\hbar \omega_{\mathbf{k}}}{e^{\beta \hbar \omega_k} - 1} \,. \tag{1.87}$$

In the continuum limit, and taking into account that $\omega_k = ck$, one can write the average number density of photons:

$$n = \frac{\langle \hat{N} \rangle}{V} = \frac{1}{\pi^2} \int_0^\infty dk\, k^2 \frac{1}{e^{\frac{\hbar c k}{k_B T}} - 1} = \frac{1}{\pi^2 c^3} \int_0^\infty d\omega \frac{\omega^2}{e^{\beta \hbar \omega} - 1} = \frac{2\zeta(3)}{\pi^2 c^3 \hbar^3 \beta^3} \,. \tag{1.88}$$

where $\zeta(x)$ is the Riemann zeta function and $\zeta(3) \simeq 1.202$. Similarly, the thermal-averaged energy density

$$\mathcal{E} = \frac{\langle \hat{H} \rangle}{V} = \frac{c\hbar}{\pi^2} \int_0^\infty dk \frac{k^3}{e^{\beta c \hbar k} - 1} = \frac{\hbar}{\pi^2 c^3} \int_0^\infty d\omega \frac{\omega^3}{e^{\beta \hbar \omega} - 1} \,. \tag{1.89}$$

By writing $\mathcal{E}$ in terms of an integral of the energy density per unit of angular frequency $\rho(\omega)$

$$\mathcal{E} = \int_0^\infty d\omega\, \rho(\omega) \,, \tag{1.90}$$

we identify

$$\rho(\omega) = \frac{\hbar}{\pi^2 c^3} \frac{\omega^3}{e^{\beta \hbar \omega} - 1}, \tag{1.91}$$

as the formula of the blackbody radiation, obtained for the first time in 1900 by Max Planck. The previous integral can be explicitly calculated and it gives

$$\mathcal{E} = \frac{\pi^2 k_B^4}{15 c^3 \hbar^3} T^4 \,, \tag{1.92}$$

which is the Stefan-Boltzmann law.

Notice that both the number density n and the energy density $\mathcal{E}$ tend to zero in the low-temperature limit. Indeed, the zero chemical potential implies that photons simply disappear when decreasing the temperature.

1.5 Gas of Massive Bosons and Bose-Einstein Condensation

The case of a gas of massive bosons presents distinctive features with respect to the case of massless bosons discussed above. In this case, the number of particles is fixed regardless of the temperature; therefore, they cannot disappear. Instead, they will arrange themselves on the available energy levels to yield a total energy in line with the corresponding temperature.

We recall that the total number of bosons is

$$N = \sum_\alpha N_\alpha = \sum_\alpha \frac{1}{e^{\beta(\epsilon_\alpha - \mu)} - 1} . \tag{1.93}$$

This suggests that at any temperature, the maximum number of bosons will be present in the ground state, a result neither new, since the same occurs in the classical Maxwell-Boltzmann distribution, nor surprising, as at positive temperatures we always expect that lower energy states are more populated than higher energy states. However, if the chemical potential is constant (and always smaller than ϵ_0, to fulfil the condition necessary to obtain Eq. (1.56)) even the ground state will become progressively depopulated in the limit of low temperatures. This is incompatible with the request of a fixed number of particles. Therefore, in order to populate at least the ground state at low temperature, we must request that the chemical potential depends on temperature in such a way that it progressively approaches the ground state in the $T \to 0$ limit. Only in this way, when $\mu \to \epsilon_0$, there is hope to compensate $\beta \to +\infty$ in Eq. (1.93) to obtain a nonzero population in the ground state. Moreover, this is the maximum value of μ allowed, as otherwise the series in Eq. (1.56) cannot be resummed. The detailed analysis now proceeds, with the task to identify the nature of this phenomenon, using again the continuum approximation. In order to discuss also a possible dependence on the effective dimensionality of the gas, we proceed by considering a generic D dimension rather than the $D = 3$ case alone.

Let us now consider a nonrelativistic Bose gas of atoms with mass m in a hyper-cubic box of volume L^D. The single-particle energy spectrum for noninteracting particles is $\epsilon_k = \hbar^2 k^2 / 2m$. The continuum limit will be achieved as in Eq. (1.68) but generalized to D dimensions:

$$\sum_{\mathbf{k}} \to L^D \int \frac{d^D \mathbf{k}}{(2\pi)^D} . \tag{1.94}$$

1.5 Gas of Massive Bosons and Bose-Einstein Condensation

However, by setting $\epsilon_0 = 0$, we cannot repeat the same procedure as for the photon gas, because otherwise we will miss the ground state which is not represented, in the low-temperature limit, using the continuum approximation. It is therefore wiser to single out the ground state from the continuum approximation, by writing the total number density $n = N/L^D$ as

$$n = \frac{1}{L^D}\frac{1}{e^{-\beta\mu}-1} + \int \frac{d^D\mathbf{k}}{(2\pi)^D}\frac{1}{e^{\beta(\frac{\hbar^2 k^2}{2m}-\mu)}-1}, \tag{1.95}$$

where the first term on the right-hand side is the density number of particles in the ground state, which we can indicate with n_0. By evaluating the integral, we obtain

$$n = n_0 + \frac{1}{\lambda_T^3} g_{D/2}(z), \tag{1.96}$$

where λ_T is the thermal de Broglie wavelength already defined in Eq. (1.78) and $g_{D/2}(z)$ is a Bose function, defined in general as

$$g_n(z) = z\frac{\int_0^\infty dx\, \frac{x^{n-1}}{z^{-1}e^x - 1}}{\int_0^\infty dx\, \frac{x^{n-1}}{z^{-1}e^x}} = \sum_{k=1}^\infty \frac{z^k}{k^n}. \tag{1.97}$$

Notice that, in analogy to the Fermi functions in Eq. (1.79), the Bose functions tend to z in the high-temperature limit, and it is always verified the inequality $f_n(z) < g_n(z)$, consistent with the more physical consideration that the average occupation number of a Bose gas is always higher than the one of the corresponding Fermi gas; see also Fig. 1.2. In the $L \to \infty$ limit and setting $z = 1$, for $D = 3$, we get

$$n = n_0 + n\left(\frac{T}{T_c}\right)^{3/2}, \tag{1.98}$$

with

$$T_c = \frac{1}{2\pi k_B \zeta(3/2)^{2/3}} \frac{\hbar^2}{m} n^{2/3}, \tag{1.99}$$

a characteristic temperature, and $\zeta(z)$ the Riemann zeta function ($\zeta(3/2) = 2.6123753\ldots$). Evidently, such a formula should hold only when $T < T_c$, as otherwise we would have $n_0 < 0$. Therefore, we find that below this temperature particles start to significantly pile up on the ground state, until they are all in the ground state at $T = 0$. Indeed, we can introduce the density of the particles in the ground state, relative to the density of the total number of particles, also called condensate fraction:

$$\frac{n_0}{n} = 1 - \left(\frac{T}{T_c}\right)^{3/2}, \tag{1.100}$$

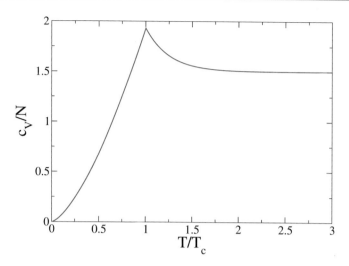

Fig. 1.4 Specific heat per atom in units of the Boltzmann constant k_B, for an ideal massive Bose gas confined in a three-dimensional box, versus the temperature T normalized to the critical temperature T_c for the onset of Bose-Einstein condensation

valid for $T < T_c$, while we have $n_0/n = 0$ for $T > T_c$. The characteristic temperature T_c separates two regions, one in which the ground state is not (or scarcely, macroscopically speaking) populated from another in which a macroscopic fraction of atoms pile up in the ground state until, at $T = 0$, all atoms are in the ground state. This is reminiscent of a phase transition, even if we have not introduced any interatomic interaction. The critical temperature of Eq. (1.99) and the condensate fraction Eq. (1.100) of noninteracting massive bosons were obtained in 1925 by Albert Einstein extending previous results derived by Satyendra Nath Bose for a gas of photons. This is the so-called Bose-Einstein condensation [6, 7].

As previously discussed, in the experiments with ultracold atomic gases, a relevant measured quantity is the specific heat, Eq. (1.82). In Fig. 1.4, we plot the specific heat C_v as a function of the temperature T for a gas of ideal bosons: its cusp behavior at the critical temperature T_c, with a discontinuity of the first derivative of the specific heat, is the signature of a phase transition. Notice that this behavior is completely different from the one of a Fermi gas in which, rather than a singular behavior at a given temperature, we observe a smooth change from the Dulong-Petit limit to the linear dependence of the Sommerfeld formula as the temperature is progressively decreased. The latter behavior is more characteristic of a crossover between two distinct regimes, rather than an abrupt change.

In the thermodynamic limit, the fugacity $z = e^{\beta\mu}$ is such that $z = 1$ for $T < T_c$, while for $T > T_c$ the fugacity z is obtained by numerically solving Eq. (1.96) with $n_0 = 0$. From the knowledge of z, we can derive the chemical potential μ, which is zero for $T < T_c$, and also the internal energy E. In Fig. 1.5, the dependence of the chemical potential for a Bose gas versus temperature is plotted and compared to the

1.5 Gas of Massive Bosons and Bose-Einstein Condensation

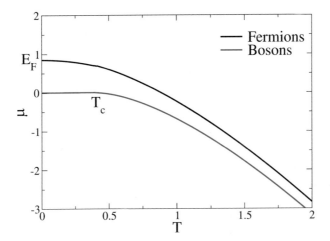

Fig. 1.5 Chemical potential for Bose-Einstein and Fermi gases, both confined in a three-dimensional box, versus temperature (axes in arbitrary units). The Fermi gas maintains a positive chemical potential at low temperature, equal to the Fermi energy at zero temperature, while in the same quantum degenerate regime, the Bose gas has its chemical potential locked to a zero value until the critical temperature for Bose-Einstein condensation T_c is reached. At higher temperatures, toward the nondegenerate, classical regime, the chemical potentials become negative and the two curves are approaching each other

case of a Fermi gas. The two curves obviously tend to overlap in the nondegenerate regime (at high temperature), while in the limit of zero temperature, they differ in the fact that the Bose gas locks its chemical potential to a zero value below T_c, while the Fermi gas reaches its Fermi energy with a parabolic dependence, as from the Sommerfeld expansion in Eq. (1.80).

It is important to realize that singling out the ground state from the other states is crucial, as in $D = 3$ this is not represented at all in the continuum limit: "No condensation without representation!". Furthermore, this phenomenon depends on dimensionality, because in the $T \to 0$ limit even the first excited states can be misrepresented in the continuum approximation. The integration over momenta involves infinitesimal volume elements dp, $p\,dp$, and $p^2\,dp$ in 1, 2, and 3 dimensions, respectively. Therefore, the higher is the dimensionality, the smaller are the chances that the few excited states (for which $p \neq 0$ but $p \to 0$ rather quickly) can contain most of the particles in the $T \to 0$ limit, without the need for them to pile up in the ground state. This suggests that in lower dimensionalities Bose-Einstein condensation may not occur.

Bose-Einstein condensation is therefore emerging from a subtle interplay between the Bose-Einstein statistics and the available density of excited states, the latter depending on the energy-wave vector dispersion relationship and the effective dimensionality. By repeating the analysis in $D = 1, 2$, we complete the possible dependence on the system dimensionality as follows:

$$k_B T_c = \begin{cases} \frac{1}{2\pi \zeta(3/2)^{2/3}} \frac{\hbar^2}{m} n^{2/3} & \text{for } D = 3 \\ 0 & \text{for } D = 2 \\ \text{no solution} & \text{for } D = 1 \ . \end{cases} \quad (1.101)$$

Therefore, the $D = 3$ case is distinctive as the pile up of particles in the ground state already occurs at a nonzero temperature. In $D = 2$, this occurs only at zero temperature (as expected also classically if $\mu \to 0$) and in $D = 1$ does not occur at all. This result was extended to other phase transitions of interacting systems [14]. The so-called Mermin-Wagner theorem, applied to our context, states that there is no Bose-Einstein condensation at finite temperature in homogeneous systems with sufficiently short-range interactions in dimensions $D \leq 2$.

It is also worth to comment on the analogies and differences between Bose-Einstein condensation and lasing. If conceived as bosonic pileup in the ground state, Bose-Einstein condensation differs from lasing in that the latter occur with a macroscopic number of photons in an arbitrary state, in the optical context called "mode," in general much higher in energy than the ground state allowed by the cavity. This difference is purely of practical character, and Bose-Einstein condensation has been also demonstrated in excited states. Still, the most natural and simple way to achieve such a state is through cooling of the atomic sample.

How the results above are robust with respect to the choice of other, more realistic, external potentials rather than an infinite well box? And, more importantly, is it possible to identify such a phase transition even in the presence of interatomic interactions? These two questions will be addressed in the next two subsections.

1.5.1 Noninteracting Bosons in a Harmonic Trap

Although the case of atoms trapped in a uniform potential with hard walls emulating an infinite well barrier has been also experimentally demonstrated, the external potential in which atoms are usually confined can be approximated, at least at the very low temperatures required for quantum degeneracy, as a harmonic potential. We then consider a bosonic gas of N noninteracting spinless particles under harmonic confinement given by the external trapping potential, assumed for simplicity isotropic:

$$U(\mathbf{r}) = \frac{1}{2} m \omega^2 (x^2 + y^2 + z^2) \ . \quad (1.102)$$

This potential is smoother than the hard wall potential and privileges a point in space, the origin, around which the particles are trapped. The associated quantum

1.5 Gas of Massive Bosons and Bose-Einstein Condensation

mechanical characteristic length, also measuring the average size of the condensate, is given by the quantity:

$$a_H = \sqrt{\frac{\hbar}{m\omega}}, \tag{1.103}$$

present in all the single-particle energy eigenstates and determining the distance from the origin on which the wave function starts decreasing exponentially to a zero value.

We want to calculate the critical temperature T_c of Bose-Einstein condensation as a function of the number N of bosons and the condensate fraction N_0/N as a function of the temperature T. One must generalize previous results taking into account the external potential $U(\mathbf{r})$ acting on bosons. Here, the semiclassical approximation with a continuum of states is translated into the condition of considering temperatures satisfying $k_B T \gg \hbar\omega$, such that the discreteness of the energy levels of the harmonic oscillator does not play a significant role. Then, we introduce the shifted single-particle energy:

$$\epsilon_k(\mathbf{r}) = \frac{\hbar^2 k^2}{2m} + U(\mathbf{r}) - \mu. \tag{1.104}$$

In this way, one can write the local number density as

$$n(\mathbf{r}) = n_0(\mathbf{r}) + \int \frac{d^3 k}{(2\pi)^3} \frac{1}{e^{\frac{\epsilon_k(\mathbf{r})}{k_B T}} - 1}, \tag{1.105}$$

where $n_0(\mathbf{r})$ is the local condensate density. Below T_c, one can set $\mu = 0$. Moreover, at T_c by definition $n_0(\mathbf{r}) = 0$ and consequently

$$n(\mathbf{r}) = \int \frac{d^3 k}{(2\pi)^3} \frac{1}{e^{\frac{\frac{\hbar^2 k^2}{2m} + U(\mathbf{r})}{k_B T_c}} - 1}. \tag{1.106}$$

After integration over wave vectors, one obtains

$$n(\mathbf{r}) = \frac{1}{\lambda_{T_c}^3} g_{3/2}(e^{-\frac{U(\mathbf{r})}{k_B T_c}}), \tag{1.107}$$

where $\lambda_{T_c} = (2\pi\hbar^2/(m k_B T_c))^{1/2}$ is the de Broglie wavelength at the critical temperature T_c and $g_n(z)$ is again the Bose function of Eq. (1.97). Moreover, integrating over coordinates, one gets

$$N = \int d^3 r \, n(\mathbf{r}) = g_3(1) \left(\frac{\hbar\omega}{k_B T_c}\right)^3, \tag{1.108}$$

from which

$$T_c = \frac{\hbar\omega}{g_3(1)\,k_B} N^{1/3} \simeq 0.94 \frac{\hbar\omega}{k_B} N^{1/3}. \tag{1.109}$$

For $T \leq T_c$, one can write

$$N = N_0 + \int \frac{d^3 r\, d^3 k}{(2\pi)^3} \frac{1}{e^{\frac{\frac{\hbar^2 k^2}{2m}+U(\mathbf{r})}{k_B T}} - 1}. \tag{1.110}$$

After integration over coordinates and wave vectors, and taking into account Eq. (1.109), we obtain

$$\frac{N_0}{N} = 1 - \left(\frac{T}{T_c}\right)^3. \tag{1.111}$$

Notice that the dependence of the condensate fraction upon the temperature is similar to the one of the homogeneous case in Eq. (1.100), differing only in the exponent of the power-law dependence.

1.5.2 Noninteracting Fermions in a Harmonic Trap

The analysis we have discussed for noninteracting bosons in an external potential $U(\mathbf{r})$ can be adapted to the case of noninteracting fermions, taking into account that the statistical distribution must be the Fermi-Dirac one. In particular, the local number density of spin-polarized fermions at temperature T is given by

$$n(\mathbf{r}) = \int \frac{d^3 k}{(2\pi)^3} \frac{1}{e^{\frac{\xi_k(\mathbf{r})}{k_B T}} + 1}, \tag{1.112}$$

where $\xi_k(\mathbf{r})$ is still given by Eq. (1.104). After integration over wave vectors, from Eq. (1.112), one obtains

$$n(\mathbf{r}) = \frac{1}{\lambda_T^3} f_{\frac{3}{2}}\!\left(e^{\frac{\mu - U(\mathbf{r})}{k_B T}}\right), \tag{1.113}$$

where λ_T is the de Broglie wavelength and $f_n(x)$ is the Fermi function, defined in Eq. (1.79).

At zero temperature, where the chemical potential μ becomes the Fermi energy ϵ_F, the density profile reads

$$n(\mathbf{r}) = \left(\frac{m}{2\pi \hbar^2}\right)^{3/2} \frac{1}{\Gamma(5/2)} (\epsilon_F - U(\mathbf{r}))^{3/2}\, \Theta(\epsilon_F - U(\mathbf{r})), \tag{1.114}$$

with $\Theta(x)$ the Heaviside step function and $\Gamma(5/2) = 3\sqrt{\pi}/4$. Integrating over the coordinates, one finds

$$T_F = 6^{1/3} \frac{\hbar\omega}{k_B} N^{1/3} \simeq 1.82 \frac{\hbar\omega}{k_B} N^{1/3}, \qquad (1.115)$$

which differs from the expression for the critical temperature of a Bose gas, Eq. (1.109), just by a numerical factor.

In both cases, T_c and T_F are directly proportional to the angular frequency of the harmonic oscillator schematizing the external trapping potential. Thus, a deeper quantum degeneracy regime can also be reached, for a given temperature, by increasing the strength of the trapping potential, a fact that is, for instance, exploited in optical lattices in which the characteristic length a_H is of the order of the light wavelength, as we will discuss in Chap. 2. In the case of anisotropic, but still harmonic, trapping, the angular frequency ω is superseded by the geometrical mean of the three frequencies along the three axes, $\omega \to \bar{\omega} = (\omega_x \omega_y \omega_z)^{1/3}$.

1.6 Interacting Condensates: Gross-Pitaevskii Equation

We now focus the attention on the more realistic case in which there are interatomic forces. Let us consider the many-body Hamiltonian operator of N identical spinless particles

$$\hat{H} = \sum_{i=1}^{N} \left(-\frac{\hbar^2}{2m} \nabla_i^2 + U(\mathbf{r}_i) \right) + \frac{1}{2} \sum_{\substack{i,j=1 \\ i \neq j}}^{N} V(\mathbf{r}_i - \mathbf{r}_j), \qquad (1.116)$$

where $U(\mathbf{r})$ is the external potential and $V(\mathbf{r} - \mathbf{r}')$ is the interatomic potential. In the case of dilute gases, we assume (Fermi pseudo-potential) that

$$V(\mathbf{r}) = g\, \delta^{(3)}(\mathbf{r}) \qquad (1.117)$$

with $\delta(\mathbf{r})$ the Dirac delta function and

$$g = \int V(\mathbf{r})\, d^3\mathbf{r}. \qquad (1.118)$$

is a constant proportional to the strength of the interatomic coupling.

The assumption of locality in the interatomic potential is justified by the fact that in ultracold and dilute atomic gases the average distance d between particles can be about 10^3 times larger than the characteristic range r_s (with $r_s \simeq 3 \times 10^{-10}$ m) of the interparticle interaction. From scattering theory (see Appendix C), the

s-wave scattering length a_s of the interatomic potential can be written, in the Born approximation, as

$$a_s = \frac{m}{4\pi\hbar^2} \int V(\mathbf{r})\, d^3\mathbf{r}, \tag{1.119}$$

and consequently, from Eqs. (1.118) and (1.119), the strength g of the contact interaction can be written in terms of a_s as

$$g = \frac{4\pi\hbar^2}{m} a_s. \tag{1.120}$$

At zero temperature, by assuming that the system remains a pure Bose-Einstein condensate despite the interparticle interaction—a reasonable assumption if the interatomic interactions are somehow small—one can write the many-body wave function as

$$\Phi(\mathbf{r}_1, \ldots, \mathbf{r}_N) = \prod_{i=1}^{N} \phi(\mathbf{r}_i), \tag{1.121}$$

which is manifestly symmetric under exchange of any two particles and where $\phi(\mathbf{r})$ is a normalized single-particle wave function

$$\int d^3\mathbf{r}\, |\phi(\mathbf{r})|^2 = 1. \tag{1.122}$$

This wave function $\phi(\mathbf{r})$ can be determined by minimizing the expectation value of the Hamiltonian operator Eq. (1.116):

$$E_N = \langle \Phi | \hat{H} | \Phi \rangle = \int d^3\mathbf{r}_1 \ldots d^3\mathbf{r}_N\, \Phi^*(\mathbf{r}_1, \ldots, \mathbf{r}_N) \hat{H} \Phi(\mathbf{r}_1, \ldots, \mathbf{r}_N). \tag{1.123}$$

After some calculations, one finds that

$$E_N = N \int d^3\mathbf{r} \left\{ \phi^*(\mathbf{r}) \left[-\frac{\hbar^2}{2m} \nabla^2 + U(\mathbf{r}) \right] \phi(\mathbf{r}) + \frac{g}{2}(N-1)|\phi(\mathbf{r})|^4 \right\}, \tag{1.124}$$

showing that E_N is a functional of $\phi(\mathbf{r})$. Typically, for many-body systems, $N \gg 1$, and then, one can approximate $N - 1 \simeq N$. In the opposite case of $N = 1$, this becomes the energy functional for a single-particle Schrödinger equation.

1.6 Interacting Condensates: Gross-Pitaevskii Equation

The minimization of E_N with respect to $\phi(\mathbf{r})$ must take into account the constraint of Eq. (1.122). Explicitly, one must minimize the functional:

$$\Omega[\phi(\mathbf{r})] = E_N[\phi(\mathbf{r})] - \mu N \int d^3\mathbf{r} |\phi(\mathbf{r})|^2, \qquad (1.125)$$

where μ is the Lagrange multiplier. It is important to stress that the Lagrange multiplier μ is the zero-temperature chemical potential of the system. In fact, the chemical potential is defined as the change in internal energy when a particle is added or removed (in this case at both zero entropy and temperature), $\mu = E_N - E_{N-1}$. On the basis of this definition, we obtain

$$\mu = \int d^3\mathbf{r} \left\{ \phi^*(\mathbf{r}) \left[-\frac{\hbar^2}{2m}\nabla^2 + U(\mathbf{r}) \right] \phi(\mathbf{r}) + g N |\phi(\mathbf{r})|^4 \right\}. \qquad (1.126)$$

This is exactly the same results one finds inserting $\phi^*(\mathbf{r})$ on the left side of the stationary equation (1.127) and integrating over $\mathbf{r}$, because of Eq. (1.122).

The minimization of Eq. (1.125) yields a time-independent equation:

$$\left[-\frac{\hbar^2}{2m}\nabla^2 + U(\mathbf{r}) + g N |\phi(\mathbf{r})|^2 \right] \phi(\mathbf{r}) = \mu\, \phi(\mathbf{r}). \qquad (1.127)$$

If $g \neq 0$, this equation is a nonlinear Schrödinger equation with cubic nonlinearity. The nonlinear term $g N |\phi(\mathbf{r})|^2$ is the mean-field potential, as seen by one boson and due to the other $N - 1 \simeq N$ bosons. Notice that, by introducing $\psi(\mathbf{r}) = \sqrt{N}\phi(\mathbf{r})$, also called macroscopic wave function or order parameter of the Bose condensate, Eq. (1.127) can be rewritten as

$$\left[-\frac{\hbar^2}{2m}\nabla^2 + U(\mathbf{r}) + g |\psi(\mathbf{r})|^2 \right] \psi(\mathbf{r}) = \mu\, \psi(\mathbf{r}). \qquad (1.128)$$

While $|\phi(\mathbf{r})|^2$ expresses the probability density for a Bose particle to be in $\mathbf{r}$, $|\psi(\mathbf{r})|^2 = n(\mathbf{r})$ represents the number density of the gas, a quantity of more direct experimental accessibility.

By extending the discussion to the time-dependent case, we obtain a time-dependent equation:

$$i\hbar \frac{\partial}{\partial t} \Psi(\mathbf{r}, t) = \left[-\frac{\hbar^2}{2m}\nabla^2 + U(\mathbf{r}) + g |\Psi(\mathbf{r}, t)|^2 \right] \Psi(\mathbf{r}, t). \qquad (1.129)$$

from which Eq. (1.128) can also be derived by separating the temporal and spatial dependence as $\Psi(\mathbf{r}, t) = \exp(-i\mu t/\hbar)\psi(\mathbf{r})$, under the hypothesis of a time-independent potential energy $U(\mathbf{r})$ and a time-independent scattering length.

Equations (1.128) and (1.129) were obtained independently in 1961 by Eugene Gross [12] and Lev Petrovich Pitaevskii [18], and named after them, in the framework of a simplified model to describe liquid ^{4}He in its low-temperature superfluid phase. However, in the case of liquid helium, the average distance d between atoms is close to the effective range r_s of the Van der Waals potential. Even if the interatomic interactions are weak for a noble gas, this makes the description of superfluid ^{4}He in terms of the Gross-Pitaevskii equation rather incomplete. This is confirmed by the finding, through neutron scattering experiments, that at very low temperature the condensate fraction in ^{4}He is only 8%, confirming the presence of substantial occupation into excited states even at the lowest explored temperatures. Instead, dilute Bose gases at very low temperature can be described reliably by Eqs. (1.128) and (1.129).

There are also limitations to the validity of the Gross-Pitaevski equation, apart from the obvious one that it relies on having a zero-temperature Bose gas. It is valid if only one mode of the condensate is occupied, therefore cannot describe states with small amounts of atoms in excited levels, the quantum depletion, or "fragmented" condensates in which there is macroscopic occupation of two or more energy levels. Finally, it does not describe the case of atomic systems exhibiting long range interatomic interactions, the so-called dipolar gases, for which Eq. (1.117) does not hold. In this case, it is still possible to generalize the Gross-Pitaevskii equation including a nonlocal interaction term.

While Eq. (1.128) can be always solved numerically with nearly arbitrary accuracy given the current computational resources, it is quite often useful to discuss more manageable analytical solutions, even if they rely on approximations, as we will discuss next.

1.6.1 Thomas-Fermi Approximation

Given the stationary Gross-Pitaevskii equation (1.128), if the nonlinear term is positive and sufficiently large, the kinetic energy density can be neglected, obtaining a simplified, purely algebraic form of the Gross-Pitaevskii equation in the so-called Thomas-Fermi approximation [9, 22, 23]:

$$\left[U(\mathbf{r}) + g|\psi(\mathbf{r})|^2 \right] \psi(\mathbf{r}) = \mu \, \psi(\mathbf{r}) \, , \qquad (1.130)$$

with solution

$$|\psi(\mathbf{r})|^2 = \begin{cases} \frac{1}{g} [\mu - U(\mathbf{r})] & \text{if } \mu - U(\mathbf{r}) \geq 0 \\ 0 & \text{if } \mu - U(\mathbf{r}) \leq 0 \end{cases} . \qquad (1.131)$$

This approximation holds for macroscopic wave functions $\psi(\mathbf{r})$ sufficiently smooth. If it is possible to identify a characteristic length scale ℓ on which the wave function varies in space, then the kinetic energy density $\simeq \hbar^2/(2m\ell^2)$ can be compared to the mean-field energy density gn. This identifies the Thomas-Fermi regime as the one for which

$$\frac{\hbar^2}{2m\ell^2} \ll gn , \tag{1.132}$$

which implies $\ell \gg \hbar/(2mgn)^{1/2}$. The quantity in the right-hand side of this inequality is also called "healing length":

$$\xi = \frac{\hbar}{(2mgn)^{1/2}}, \tag{1.133}$$

and, apart from a numerical factor, is the product Na_s. The healing length represents the length scale on which or above which a sharp "scar" of size ℓ made on the condensate profile spending an energy $\simeq \hbar^2/2m\ell^2$ can be cured at the price of using a comparable mean-field interaction energy.

Under the Thomas-Fermi approximation, the probability density is determined by the external potential. For instance, if the external potential $U(\mathbf{r})$ is an isotropic harmonic potential as in Eq. (1.102), the probability density profile is an inverted parabola truncated at the extremes of the classical motion determined by the chemical potential μ:

$$n(\mathbf{r}) = |\psi(\mathbf{r})|^2 = \begin{cases} \frac{1}{g}\left(\mu - \frac{1}{2}m\omega^2 r^2\right) & \text{if } \mu - U(\mathbf{r}) \geq 0 \\ 0 & \text{if } \mu - U(\mathbf{r}) \leq 0 \end{cases} \tag{1.134}$$

with $r = \sqrt{x^2 + y^2 + z^2}$. We can identify the radius of the condensate as the distance R_{TF} from the origin of minimum of the trapping potential $U(R_{TF}) = \mu$, the Thomas-Fermi radius. In the case of the harmonic trapping, we are considering $R_{TF} = [2\mu/(m\omega^2)]^{1/2}$.

Moreover, in this case, by using Eq. (1.122), we obtain an explicit formula for the chemical potential:

$$\mu = \frac{\hbar\omega}{2}\left(\frac{15Na_s}{a_H}\right)^{2/5}, \tag{1.135}$$

as a function of the number N of bosons, the s-wave scattering length a_s, and the characteristic length a_H of the harmonic confinement. Considering that the chemical potential for an ideal ($a_s = 0$) Bose gas in a harmonic trap is $\mu = \hbar\omega/2$, the Thomas-Fermi approximation is valid if $Na_s \gg a_H$. Since the chemical potential determines the size of the condensate in the trap, in the case of harmonic trapping, this implies sizes of the condensate much larger than the ones expected on the basis of a single-particle trapping, which is advantageous from the imaging standpoint.

The Thomas-Fermi approximation finds broad use in the analysis of experimental results, especially for fitting density profiles, using the macroscopic wave function

$|\psi(\mathbf{r})|^2 = N|\phi(\mathbf{r})|^2$. Its successful application is related to the fact that, even if g is small for a weakly interacting Bose gas, the cumulative many-body effect for large N makes the interaction energy term overwhelming with respect to the kinetic energy term, a situation typical unless condensates with very small number of atoms are considered. Its failure is also expected around the region of space for which $r \simeq R_{TF}$, in which the first spatial derivatives of the density profile of the condensate should have a discontinuity according to Eq. (1.131). In reality, we expect a smooth decay of the condensate density mantaining nonzero values even at $r > R_{TF}$, even more so if one consider finite temperature effects.

1.6.2 Gaussian Variational Approach

Approximate solutions to the Gross-Pitaevskii equation are also available by using variational methods [3, 13, 15, 21]. As an example, the Gross-Pitaevskii functional can be used to calculate the energy per particle of a Bose-Einstein condensate under harmonic confinement, as in Eq. (1.102), using a specific class of trial wave functions. In this case, it is natural to use variational wave functions of Gaussian form:

$$\phi(\mathbf{r}) = \frac{1}{\pi^{3/4} a_H^{3/2} \sigma^{3/2}} e^{-(x^2+y^2+z^2)/(2a_H^2 \sigma^2)} , \qquad (1.136)$$

where a_H is the characteristic length of the harmonic confinement and σ is a dimensionless variational parameter. Inserting this wave function into the Gross-Pitaevskii energy functional

$$E_N = N \int d^3\mathbf{r} \left\{ \phi^*(\mathbf{r}) \left[-\frac{\hbar^2}{2m} \nabla^2 + \frac{1}{2} m\omega^2 (x^2 + y^2 + z^2) \right] \phi(\mathbf{r}) + \frac{g}{2} N |\phi(\mathbf{r})|^4 \right\} , \qquad (1.137)$$

after integration, we obtain the energy E_N as a function of σ, namely,

$$E_N = N\hbar\omega \left(\frac{3}{4\sigma^2} + \frac{3}{4}\sigma^2 + \frac{\gamma}{2\sigma^3} \right) , \qquad (1.138)$$

where $\gamma = Na_s/(\sqrt{2\pi} a_H)$ is a dimensionless parameter proportional to the ratio between the interparticle energy and the kinetic energy. By extremizing the energy function, we find the optimized value of σ through the identity:

$$\sigma(\sigma^4 - 1) = \gamma . \qquad (1.139)$$

If $\gamma > 0$, $|\sigma|$ grows as $\sigma \simeq \gamma^{1/5}$ in the limit of $\sigma \to \infty$, and as $\sigma \simeq -\gamma$ if $\gamma \ll 1$. Instead, if $\gamma < 0$, there are two possible values of σ: one value corresponds to a minimum of the energy E, a metastable solution, and the other corresponds to a maximum of the energy E, an unstable solution. These two solutions exists only if $\gamma > -4/5^{5/4}$. Therefore, we expect instability of the Bose-Einstein condensate for a negative scattering length and above a given threshold for the number of atoms.

The instability of the atomic Bose-Einstein condensate in a harmonic confinement was observed experimentally in several laboratories, starting with the case of ^{7}Li [2,19,20]. At a fixed negative value of the scattering length a_s, a critical number of atoms exists, $N_c \simeq a_H/|a_s|$, usually of order of 10^3–10^4 atoms, above which there is the so-called collapse of the condensate. More flexibility in studying this effect is available by changing the scattering length using the so-called Feshbach resonances [10], as we will discuss later. This collapse corresponds to a sudden implosion of the bosonic cloud of atoms with a subsequent explosion due to a very fast release of untrapped molecules, which are formed due to three-body recombination processes [5]. Such a phenomenon resembles what happens in a supernova event, therefore is also known as Bosenova.

The variational method, at least if based on the use of Gaussian wave functions, is complementary to the Thomas-Fermi method, as it is meaningful when the kinetic term of the energy in the Gross-Pitaevskii functional is dominant with respect to the mean-field term, as expected in the small number of atom regime or in the case of a small elastic scattering length.

1.7 Problems

1.1. Write the density matrix for a two-level system characterized by the probability 1/3 of being in state $|0\rangle$ and probability 2/3 of being in state $|1\rangle$, assuming the representation of the density operator in which the two states form an orthonormal basis. Then, write the density matrix for the same system but kept in the state:

$$|\psi\rangle = \frac{1}{\sqrt{3}}|0\rangle + \sqrt{\frac{2}{3}}|1\rangle. \tag{1.140}$$

1.2. Consider the eight thermodynamical potentials having as independent variables any combination of three quantities such as entropy S, volume V, number of particles N, or their conjugate quantities temperature T, pressure p, and chemical potential μ, defined in Eq. (1.17). A quantity is called *intensive* if it does not depend on the amount of its constituents, i.e., is $O(N^0)$. If a quantity is linearly dependent on the amount of constituents, $O(N^1)$, it is instead called *extensive*.

(a) Determine which quantities among the six above are extensive or intensive.
(b) Prove that the thermodynamical potential having as dependent variables T, p, and N, called $G(T, p, N)$, must necessarily satisfy $G = \mu N$.
(c) Prove that the grand canonical potential $\Omega(T, V, \mu)$ must necessarily satisfy $\Omega = -pV$.

1.3. Show that in an intermediate regime of quantum degeneracy the average number of particles in the continuum is well approximated by averaging the number of particles for the corresponding Bose and Fermi-Dirac statistics $(\langle n_B \rangle + \langle n_F \rangle)/2$. Compare to Fig. 1.2.

1.4. Consider systems of N bosons and fermions as the ones in the ground states depicted in Fig. 1.1. Let the energy spacing between the levels be $\hbar\omega$, with ω the angular frequency of the harmonic potential in which they are trapped. Show that the chemical potential of the Bose gas is equal to the ground state energy, i.e., $\mu_B = \epsilon_0 = \hbar\omega/2$, while the chemical potential of the Fermi gas is $\mu_F = (N+1)\hbar\omega$. Now, consider the same systems at finite temperature, and provide arguments, this time using the definition of chemical potential in terms of the Helmoltz free energy, to justify why, in the high-temperature limit, the chemical potential must be necessarily negative.

1.5. Show that a two-dimensional ideal Bose gas confined in a square does not undergo Bose-Einstein condensation. Investigate the case of the same gas confined in a harmonic trap.

1.6. Calculate the Fermi energy and the internal energy for a system of N noninteracting spin 3/2 particles trapped in a harmonic potential in D dimensions, with D = 1, 2, 3.

1.8 Further Reading

- K. Huang, *Statistical Mechanics*, Wiley, 2nd edition (John Wiley & sons, New York, 1987).
- A. S. Parkins and D. F. Walls, *The physics of trapped dilute-gas Bose-Einstein condensates*, Phys. Rep. **303**, 1 (1998).
- D. A. R. Dalvit, J. Frastai, and I. D. Lawrie, *Problems on Statistical Mechanics* (IOP, Bristol and Philadelphia, 1999).
- F. Dalfovo, S. Giorgini, L.P. Pitaevskii, and S. Stringari, *Theory of Bose-Einstein condensation in trapped gases*, Review of Modern Physics **71**, 463 (1999).
- C. J. Pethick and H. Smith, *Bose–Einstein Condensation in Dilute Gases* (Cambridge University Press, Cambridge, 2001).
- M. Kardar, *Statistical Physics of Particles* (Cambridge University Press, 2007).
- S. Giorgini, L.P. Pitaevskii, and S. Stringari, *Theory of ultracold atomic Fermi gases*, Reviews of Modern Physics **80**, 1215 (2008).
- R. K. Pathria and P. D. Beale, *Statistical Mechanics*, 4th edition (Academic Press, 2021).

References

1. Bose, S.N.: Plancks Gesetz und Lichtquantenhypothese Z. Physik **26**, 178 (1924)
2. Bradley, C.C., Sackett, C.A., Hulet, R.G.: Bose-Einstein condensation of lithium: observation of limited condensate number. Phys. Rev. Lett. **78**, 985 (1997)
3. Chiofalo, M.L., Tosi, M.P.: Output from Bose condensates in tunnel arrays: the role of mean-field interactions and of transverse confinement. Phys. Lett. A **268**, 406 (2000)
4. Dirac, P.A.M.: On the theory of Quantum mechanics. Proc. Royal Society A **112**, 661 (1926)
5. Donley, E.A., Claussen, N.R., Cornish, S.L., Roberts, J.L., Cornell, E.A., Wieman, C.E.: Dynamics of collapsing and exploding Bose-Einstein condensates. Nature **412**, 295 (2001)
6. Einstein, A.: Quantentheorie der Einatomigen idealen Gases. Sitzungsber. Kgl. Preuss. Akad. Wiss. **1924**, 261 (1924)
7. Einstein, A.: Quantentheorie der Einatomigen idealen Gases. Zweite abhandlung. Sitzungsber. Kgl. Preuss. Akad. Wiss. **1925**, 3 (1925)
8. Fermi, E.: Sulla quantizzazione del gas perfetto monoatomico. Rend. Lincei **3**, 145 (1926)
9. Fermi, E.: Bestimmung einiger Eigenschaften des Atoms und ihre Anwendung auf die Theorie des periodischen Systems der Elemente. Z. Phys. **48**, 73 (1928)
10. Feshbach, H.: A unified theory of nuclear interactions. II. Ann. Phys. (N.Y.) **19**, 287 (1962)
11. Gaunt, A.L., Schmidutz, T.F., Gotlibovych, I., Smith, R.P., Hadzibabic, Z.: Bose-Einstein condensation of atoms in a uniform potential. Phys. Rev. Lett. **110**, 200406 (2013)
12. Gross, E.P.: Structure of a quantized vortex in boson systems. Nuovo Cimento **20**, 454 (1961)
13. Jackson, A.D., Kavoulakis, G.M., Pethick, C.J.: Solitary waves in clouds of Bose-Einstein condensed atoms. Phys. Rev. A **58**, 2417 (1998)
14. Mermin, N.D., Wagner, H.: Absence of ferromagnetism or antiferromagnetism in one- or two-dimensional isotropic Heisenberg models. Phys. Rev. Lett. **17**, 1133 (1966)
15. Olshanii, M.: Atomic scattering in the presence of an external confinement and a gas of impenetrable bosons. Phys. Rev. Lett. **81**, 938 (1998)
16. Pauli, W.: Über den Zusammenhang des Abschlusses der Elektronengruppen im Atom mit der Komplexstruktur der Spektren. Z. Physik **31**, 765 (1925)
17. Pauli, W.: Exclusion principle and quantum mechanics. In: Nobel Lecture, pp. 27–43 (1946)
18. Pitaevskii, L.P.: Vortex lines in an imperfect Bose gas. Zh. Eksp. Teor. Fiz. **40**, 646 (1961) [Sov. Phys. JETP 13, 451 (1961)]
19. Sackett, C.A., Stoof, H.T.C., Hulet, R.G.: Growth and collapse of a Bose-Einstein condensate with attractive interactions. Phys. Rev. Lett. **10**, 2031 (1997)
20. Sackett, C.A., Gerton, J.M., Welling, M., Hulet, R.G.: Measurements of collective collapse in a Bose-Einstein condensate with attractive interactions. Phys. Rev. Lett. **82**, 876 (1999)
21. Salasnich, L., Parola, A., Reatto, L.: Effective wave equations for the dynamics of cigar-shaped and disk-shaped Bose condensates. Phys. Rev. A **65**, 043614 (2002)
22. Schuck, P., Vinas, X.: Thomas-Fermi approximation for Bose-Einstein condensates in traps. Phys. Rev. A **61**, 043603 (2000)
23. Thomas, L.H.: The calculation of atomic fields. Proc. Cambridge Philos. Soc. **23**, 542 (1926)

Trapping and Cooling of Atoms

In this chapter, we introduce the basics for confining and cooling samples of atoms at the gaseous state. First, we discuss the need of very low temperatures for achieving quantum degeneracy in the presence of concomitant effects due to several sources of atomic interactions. The latter are more prominent at high density, therefore setting an upper bound on both density and temperature of the atomic gas. Then, we remind some notions of atomic physics necessary to understand the mechanical action of light on atoms, the radiation pressure resulting in their slowing and cooling. We then describe two classes of purely conservative traps based on the presence of magnetic and electric dipole moments in the atoms and the related cooling mechanism based on forced evaporation. Particular care is taken to discuss the limits of each trapping and cooling technique, as this has played and still plays a major role in determining optimal experimental configurations for achieving quantum degeneracy.

2.1 The Temperature-Density Window for Quantum Degeneracy

As mentioned in the introduction of the previous chapter, phenomena related to quantum degeneracy such as Bose-Einstein condensation can be masked by competing processes. For instance, if the density of the atomic gas is too large, there is a nonnegligible probability for the progressive formation of molecules, clusters, and then solids, starting with three-body interactions. Also, if the vacuum in the trapping region is not low enough, the atomic sample will occur prohibitive heating due to the atoms of the residual atmospheric gases, mainly nitrogen and oxygen, which moreover are at room temperature. Dipolar relaxation, with coupling between spin and angular momenta in two-body processes, is another source of losses. In general, these processes are stationary, and eventually, they will prevail on any dynamics related to quantum degeneracy.

This implies that studies of genuine quantum phenomena for ultracold atoms such as superfluidity features can only occur in a transient regime and outside the typical framework of a static phase diagram at thermodynamical equilibrium. In this sense, they remind us of supercooling, in which a fast cooling of a liquid temporarily maintains the sample in the liquid state even if this should correspond to a point in the phase diagram characteristic, at equilibrium, for the solid phase. This consideration affects a whole class of phenomena. For instance, the so-called "persistent" currents in atomic gases do not share stationary, long-time behavior characteristic of liquid helium or superconductors, and atoms in optical lattices do not live for the long, often indefinite times as usual crystals in solid state physics. However, their dynamics can be usually investigated in much faster timescales than for condensed matter systems, partially compensating this drawback.

Each atom maintains its individuality as far as the De Broglie wavelength associated to its motion is much smaller than the average distance from other atoms. This ensures a sort of "geometrical optics" setting in which each atom behaves independently, in the nondegenerate regime. For an ensemble of atoms at equilibrium at a given temperature, the de Broglie wavelength is determined by the kinetic energy associated to the thermal energy. The colder the sample, the larger the de Broglie wavelength. At a fixed atomic density, this means that sooner or later, the temperature will be cold enough to make the de Broglie wavelength comparable, or even larger, than the average distance between the atoms, which can be estimated as the inverse of the cubic root of the atomic density. Under this condition, the atoms lose their individuality and their corpuscular behavior, entering in a regime similar to "physical optics" in which interference and diffraction phenomena may occur. Notice that one can also reason in the complementary way, keeping the de Broglie wavelength constant, while keeping constant the temperature, and progressively reducing the distance between the atoms via an isothermal compression of the sample. Either way, the wavelike nature of the atoms will be manifest in their motion, originating what we call a quantum degenerate gas. For N atoms of mass m at equilibrium at temperature T and confined in a box of volume V (such that they have a uniform density $n = N/V$), their quadratic average velocity can be estimated as $\langle v^2 \rangle \simeq (k_B T/m)^{1/2}$. Then, the estimate for the de Broglie wavelength due to thermal motion is $\lambda_{th} \simeq h/(m k_B T)^{1/2}$, and the condition for quantum degeneracy translates into $\lambda_{th} \geq n^{-1/3}$. In this way, simply inverting the last equation, we can recover the characteristic temperatures/energies for the onset of quantum degeneracy in Eqs. (1.71) and (1.99) for Fermi and Bose gases, respectively, apart from a numerical factor of order unity. Indeed, the condition for quantum degeneracy can also be expressed in terms of a condition on a dimensionless quantity named *phase space density*, defined as $D = n\lambda_{th}^3$, the average number of particles in a cube of size equal to the thermal De Broglie wavelength. In terms of what discussed in the previous chapter for a uniformly confined Bose gas, if $D > \zeta(3/2) \simeq 2.612$, Bose-Einstein condensation occurs.

These qualitative considerations are important because they show that it is not enough to just cool atoms to low temperature to reach quantum degeneracy. The path to quantum degeneracy requires achieving both high density and low temperature.

The concrete realization of this quantum regime depends, as discussed in the previous chapter, upon the fermionic or bosonic nature of the particles. Although the details of the behavior of the translational degrees of freedom are different in the quantum degenerate regime, several features are common to both classes of atoms, at least because the precooling stages in the classical regime are rather insensitive to quantum statistics.

Having this in mind, we can estimate the minimum temperature required to observe quantum degeneracy based on the assumptions that interatomic forces start to play a role when atoms are separated by less than 100 nm and uniform confinement. This range corresponds to a maximum allowed density of 10^{15} atoms/cm^3. Recalling that ^{4}He, with density of order 10^{22} atoms/cm^3 characteristic of liquid and solid state, becomes superfluid at temperatures of order 1 K, and considering that the temperature required for quantum degeneracy scales with density, in the case of uniform confinement, through a 2/3 power law, we need to reach temperatures of order 100 μK or less to explore quantum degeneracy at the gaseous state. This of course also depends on the atomic species due to the mass dependence for the onset of quantum degeneracy, and, for instance, in the case of ^{133}Cs, this implies sub-μK temperatures. In this regard, it seems convenient to investigate hydrogen due to its smaller mass and, as a possible bonus, its overall simplicity and complete theoretical characterization. Based on this semiquantitative analysis, we rule out the use of cryogenic techniques, with liquid He as a buffer gas, to cool atomic samples to the full quantum degeneracy. Nevertheless, cryogenic techniques were initially used as a precooling stage for hydrogen and more recently have also become successfully used for cooling molecules.

2.2 Reminder of Atomic Physics

Atoms may be schematized as bound and electrically neutral systems made of electrons and a highly localized (within order of 10^{-15} m) aggregate of positively charged protons, with the neutrons playing a minor, but existing, role in atomic spectroscopy, but determining the bosonic or fermionic character of different isotopes. We can classify the atomic degrees of freedom as internal, if related to the arrangement of electrons in various bound states, and external, if related to the motion of the atoms as whole entities. Internal degrees of freedom are usually described in terms of energy levels for the electrons surrounding the nucleus, starting from the ground state. Transitions between different energy levels that can be induced by absorption and emission of photons and also occur spontaneously due to the quantum fluctuations of the electromagnetic field. External degrees of freedom are the location in space of each atom (for instance, of their nuclei) and their momenta. Usually, these external degrees of freedom are considered classical, but the distinctive feature of ultracold atom physics is that also external degrees of freedom need to be treated quantum mechanically.

The starting point to control internal and external degrees of atoms is to understand the spectroscopy of the ground state and the first excited states. As usual in physics, it is important to first consider the simplest entities, in this case hydrogen. This has also the advantage that the temperature for the onset of Bose-Einstein condensation is the largest, due to the smaller mass. Unfortunately, the absence of high-intensity lasers for the atomic transitions from and to the ground state of hydrogen has hindered its use in ultracold physics for some time, as the corresponding wavelengths fall into the ultraviolet range. Next to hydrogen, in terms of spectroscopic simplicity are the alkali-metal atoms. Due to the partial screening of the less bound electron due to the innermost electrons, the spectroscopy of these atoms falls in the visible and near infrared, where lasers with sufficient intensity were already available when this research direction was initiated.

The bosonic or fermionic character of each alkali-metal atom will be determined by the number of neutrons in the nuclei, due to the electrical neutrality of the atoms. To be a boson, an alkali-metal atom should have an even number of neutrons, vice versa for an alkali-metal atom behaving as a fermion. For instance, ^{7}Li is a boson, containing 3 protons, 4 neutrons, and 3 electrons, thus with total integer spin. Conversely, ^{6}Li is a fermion, due to the odd (3) number of neutrons. Due to nuclear pairing, the only stable (or radioactive with very long lifetime) alkali-metal atoms with odd number of neutrons are ^{6}Li and ^{40}K, strongly limiting the inventory for available fermions with hydrogen-like spectroscopy.

Always to keep matters simple, we will consider just two energy levels as close as possible to realize a two-level system. This requires a careful analysis of the energy levels based upon the strength of the transitions and the selection rules for emission and absorption of photons. Although not significantly contributing to energy shifts in themselves with respect to the gross and fine structure, perhaps surprisingly, this requires the analysis of the much smaller nucleus-electron interaction. This may be discussed in close analogy to the treatment of the fine structure due to the spin-orbit interaction in which the electron has a magnetic moment

$$\boldsymbol{\mu}_e = -g_e \mu_B \mathbf{S}, \tag{2.1}$$

where $\mu_B = e\hbar/(2m_e) = 9.274 \times 10^{-24}$ J/T is the Bohr magneton, $\mathbf{S}$ the spin operator, and g_e the electron gyromagnetic factor, usually approximated in atomic physics, by neglecting radiative corrections due to quantum vacuum, as $g_e \simeq 2$.

In the case of nuclei, we introduce the nuclear magneton μ_N by simply replacing the mass of the electron with the mass of the proton

$$\mu_N = \frac{e\hbar}{2m_p c} = \frac{\mu_B}{1836} = 1.410 \times 10^{-26} \text{J/T}, \tag{2.2}$$

and introducing a nuclear g-factor, g_I, such that the nuclear magnetic moment is

$$\boldsymbol{\mu} = g_I \mu_N \mathbf{I}, \tag{2.3}$$

where $\mathbf{I}$ is the total angular momentum operator of the nucleus. The fact that protons and neutrons are not elementary particles is corroborated by their non-

2.2 Reminder of Atomic Physics

integer g-factors, $g_p = +5.586$ and $g_n = -3.826$, respectively. Notice that, in spite of the larger g-factors of protons and neutrons, nuclear magnetic moments are suppressed by three orders of magnitude with respect to the electron counterpart. This results in hyperfine energy splittings smaller, by a factor $\simeq m_e/m_p$, with respect to the fine energy splittings, themselves in turn smaller by a factor $\simeq \alpha^2$, with $\alpha = e^2/(4\pi\hbar c) \simeq 1/137$ the fine structure constant, with respect to the Bohr energy differences, having hydrogen as a simple benchmark.

The following steps closely mimic what is already discussed for the usual spin-orbit coupling and are valid for hydrogen-like atoms. The total Hamiltonian must include a term due to the hyperfine interaction between the nuclear magnetic moment and the magnetic field felt by the nucleus due to the electron:

$$H_{\text{hf}} = -\boldsymbol{\mu} \cdot \mathbf{B_J}, \tag{2.4}$$

where the magnetic field gets two contributions, due to the angular momentum L and the spin S of the electron, $\mathbf{B_J} = \mathbf{B_L} + \mathbf{B_S}$. These two contributions can be evaluated both with a classical and a quantum treatment, leading to an expression for the magnetic field felt by the proton, even in the case of an electron in the s-state, interpreted as the magnetization due to a spherically symmetric cloud in close proximity of the nucleus. The hyperfine Hamiltonian can be written as

$$H_{\text{hf}} = \hbar \omega_{\text{hf}} \mathbf{I} \cdot \mathbf{J}, \tag{2.5}$$

where we have introduced a hyperfine coupling constant ω_{hf} with the dimensions of an angular frequency. Equation (2.5) suggests the introduction of a new operator, the total angular momentum of the atom (including the angular momentum of the nucleus) as

$$\mathbf{F} = \mathbf{J} + \mathbf{I}, \tag{2.6}$$

such that, using $\mathbf{F}^2 = (\mathbf{J} + \mathbf{I})^2$, we obtain

$$\langle \mathbf{I} \cdot \mathbf{J} \rangle = \frac{1}{2}[F(F+1) - J(J+1) - I(I+1)], \tag{2.7}$$

where the expectation value is calculated over an atomic state specified by its eigenvalues I and J. In the case of hydrogen, the calculation can be carried over analytically by using perturbation theory and leads to hyperfine energies:

$$E_{\text{hf}} = \frac{g_e g_N \mu_B \mu_N}{a_0^3 n^3 (2\ell+1)} \frac{F(F+1) - J(J+1) - I(I+1)}{J(J+1)}. \tag{2.8}$$

Selection rules for photon transitions are similar to the case of fine spectroscopy, $\Delta F = 0, \pm 1$, and the estimate of the hyperfine energy yields energy splittings

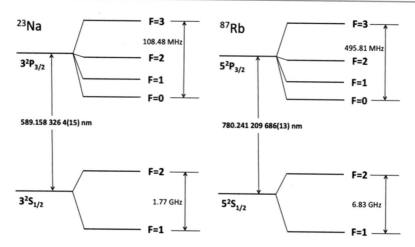

Fig. 2.1 Energy levels relevant for the D_2 line in bosonic ^{23}Na (left) and ^{87}Rb (right), the first two species that have been Bose condensed in 1995 (for more details on the properties of ^{23}Na and ^{87}Rb, see [42] and [41], respectively)

corresponding to frequencies in the GHz range. In the case of alkali metals in the ground state $2S_{1/2}$, we have $I = 3/2$ and $J = 1/2$; therefore, $F = |I - j| = 1$ and $F = I + J = 2$, two states separated by a few GHz. The $2P_{1/2} \to 2S_{1/2}$ is called D_1 transition, the $2P_{3/2} \to 2S_{1/2}$ D_2 transition. Their splitting was first observed in sodium spectra and corresponds to 589.5924 nm and 588.9950 nm, respectively. The D_2 line has an intensity twice the D_1 line, and they dominate over all other transition, determining the predominantly yellow light emission from a sodium lamp, with the D doublet easily visible even with modest resolution spectrometers. This is the reason for which the energy levels corresponding to the D_2 line are usually preferred as two-level system. In Figs. 2.1 and 2.2, we show the energy levels for the D_2 line transitions of the two bosonic isotopes of alkali atoms first condensed in 1995, ^{87}Rb and ^{23}Na, and for the two fermionic isotopes of alkali studied, respectively. As commented earlier, the choice in the latter case is limited to these two isotopes for alkali metals for stable isotopes, with ^{40}K unstable but with such a low radioactive decay rate (lifetime $\tau = 1.28 \times 10^9$ y) to consider it effectively stable in our context. In Table 2.1, we list various properties of these atoms, in particular the value of the nuclear spin I in the ground state, and the wavelengths and lifetime of the D_2 transitions.

In the presence of an external weak magnetic field, these levels will split into $2F + 1$ sublevels with different third component m_F and selection rules for m_F as $\Delta m_F = 0, \pm 1$. The $F = 1$ and the $F = 2$ states will then split into three and five Zeeman substates, respectively, and this also holds for the excited state $2P_{1/2}$, with the same value of J. The excited state $2P_{3/2}$ has $F = 0, 1, 2, 3$ and therefore a total of 18 Zeeman sublevels. The assumption of weak magnetic fields is usually valid for cold atomic clouds in the minimum of a magnetic field and corresponds to the dominance of the internal magnetic interaction between the electron and the nucleus

2.2 Reminder of Atomic Physics

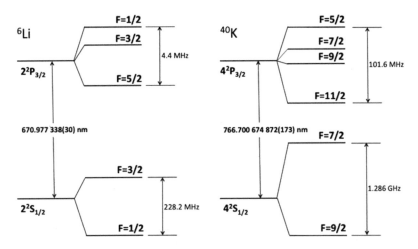

Fig. 2.2 Energy levels relevant for the D_2 line of fermionic alkali-metal isotopes, the widespread ^{6}Li (left), and ^{40}K (right), the first two species brought to a temperature smaller than the Fermi temperature in 1999 (for more details on the properties of ^{6}Li and ^{40}K, see [14] and [44], respectively)

with respect to the interaction of the electron with the external magnetic field. In the opposite case, the classification of the states in terms of F is no longer valid, the Paschen-Bach effect, and energy levels as well as the selection rules are determined by J and m_J.

In selecting the energy levels for the atomic manipulation, one has to keep in mind the small difference between the hyperfine energies. In particular, if two states are chosen as playing the privileged role of a "two-level" system, it is important to be sure that the electron does not fall into adjacent energy levels when making a transition. The probability to make such inappropriate, for our goal, transitions may be nonnegligible considering the typical number of cyclic transitions between two energy levels for laser cooling. In this case, it is necessary to provide auxiliary laser beams to repump the atoms in the desired energy levels. Due to the small frequency shifts in hyperfine transitions, these auxiliary beams are actually provided by the same laser, using devices able to shift the laser frequency by the required amount, order of few GHz at most. These devices are called electro-optical modulators (EOMs) if their operating mechanism is based on the electro-optic effect, in which the laser beam crosses a nonlinear crystal in which an electric field creates a modulation of one of the parameters (frequency in this case, but also amplitude, polarization, or phase possibly). Alternatively, they are called acousto-optical modulators (AOMs) if instead they are based on the acousto-optic effect, with a piezoelectric transducer creating, upon application of a variable electric field, stationary sound waves inside a crystal. The sound waves modulate the density of the crystal and therefore the index of refraction. As a consequence, the primary laser beam at the output of the crystal will also contain harmonics at the sum

Table 2.1 Properties of ^{6}Li, ^{23}Na, ^{40}K, and ^{87}Rb of interest for laser cooling. For each species, we report the mass in atomic units, the melting point temperature (MPT), the nuclear spin I (with all alkali metals having total electronic spin $S=1/2$), the wavelength of the D$_2$ line, the spontaneous emission lifetime, the Doppler temperature, and the recoil temperature

Atom	Mass (u)	MPT (°C)	I	λ D2-line (nm)	τ (ns)	T_D (μK)	T_{rec} (μK)
^{6}Li	6.015	180.54	1	670.977	27.102	141	3.536
^{23}Na	22.989	97.80	3/2	589.158	16.249	235	2.340
^{40}K	39.964	774.0	4	766.701	26.37	145	0.404
^{87}Rb	86.909	688	3/2	780.241	26.24	146	0.362

and difference of multiples of the frequency of the oscillating electric field, with a corresponding angular deflection. This last also allow for fast deflections of a laser beam, an advantage used to arbitrarily generate trapping geometries as we will discuss in the section devoted to optical dipole traps.

2.3 Doppler Cooling

A cooling mechanism for atoms may rely on the existence of forces of dissipative character due to the detuning between some selected and dominant, in terms of rates, atomic transitions and the frequency of photons impinging on the atoms. Photons, when absorbed by atoms, deliver both their energy and momentum. This is the microscopic counterpart of the radiation pressure demonstrated with macroscopic masses by Lebedev [24] and, independently, by Nichols and Hull [34] at the beginning of the twentieth century. In 1933, Frisch observed the lateral deflection of a sodium atomic beam due to light from a sodium lamp sent perpendicular to the atomic beam [11]. It took another four decades, and the development of lasers, to propose the use of radiation pressure from light for slowing and cooling atoms, mainly for the goal of high-resolution spectroscopy [3, 18, 25–27].

If a stream of photons is absorbed by an atom, the latter will move in the original direction of the photons to conserve energy and momentum. While the atom absorbs momentum in the same direction of the absorbed photons, its de-excitation due to spontaneous emission has instead a random character and will occur in all directions. This means that the atoms will be slowed down along the original direction of their motion while also developing a random motion in the two directions orthogonal to the light beam. The rate of momentum change, i.e., the effective average force exerted on the atom, will depend on the energy of the photons and on the absorption probability. The latter is maximized if the photon energy is equal to the energy difference between the two selected atomic levels, when it is "in resonance." This dynamics is illustrated in part (a) of Fig. 2.3, in which an absorption event of a photon at time t_0 for an atom in the ground state transfers the entire photon momentum (on top of energy and angular momentum) to the atom, here imagined excited in a p state at time t_1 (as in the case of the selection rules for hydrogen), then followed by the spontaneous emission of a photon in a

2.3 Doppler Cooling

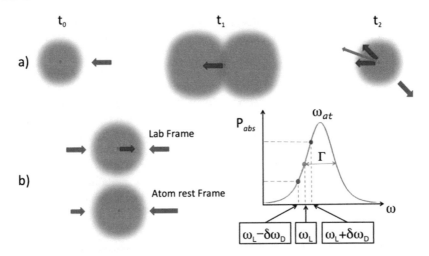

Fig. 2.3 Radiation pressure on atoms arising from absorption and spontaneous emission of photons in the case of (**a**) a single laser beam tuned at the atomic resonance and (**b**) two counterpropagating beams red-detuned with respect to the atomic resonance and near the leftmost inflection point of the absorption curve. Due to the randomness of the spontaneous emission, the entropy of the atom-photon system is increased in time. The created photon has a definite modulus of its momentum, but a random direction, unlike the initial photon. In the simplest case, (**a**) the atom at time t_2 has a momentum with a random direction, and depending on this direction, the modulus of its momentum varies continuously between zero (when the photon is emitted horizontally toward the left) and twice the recoil momentum (when the photon is emitted horizontally toward the right)

random direction at a random time t_2 depending on the lifetime of the transition between the two energy levels. The direction of emission of the photon at t_2 is also random, and repeated absorption events will result in a sustained recoil of the atom linearly dependent on the number of absorption. If the atom has an initial velocity opposite to the one of the photons, this will result in its slowing in the horizontal direction, with a diffusion process in the two orthogonal directions. It is important to notice that the photon will be absorbed even if its energy is not coinciding with the energy difference between the two atomic levels, in particular even if its energy is smaller, provided it is within a few linewidths from resonance.

Let us consider an atom schematized as a two-level system (with the energy difference between the two levels $E_2 - E_1 = \Delta E = \hbar\omega_{\text{at}}$, and $\Gamma = 1/\tau$ the linewidth of the transition, the reciprocal of the lifetime for spontaneous emission). Let us also assume that the atom is moving along the x-axis with velocity $\mathbf{v}$ in the presence of an electromagnetic plane wave with wave vector $\mathbf{k}$ and angular frequency $\omega_L = ck$, with a detuning $\delta\omega = \omega_L - \omega_{\text{at}}$. The average force on the atom due to the absorption and spontaneous emission of photons will be written as

$$\langle \mathbf{F} \rangle_{\text{slow}} = \hbar\mathbf{k}\frac{\Gamma}{2}\frac{I/I_0}{1 + I/I_0 + [2(\delta\omega - \mathbf{k}\cdot\mathbf{v})/\Gamma]^2} = \hbar\mathbf{k}R, \quad (2.9)$$

where $I_0 = \hbar\omega_{at}^3\Gamma/(12\pi c^2)$ is the saturation intensity, and we have implicitly introduced the photon scattering rate R as relating the average force to the photon momentum $\hbar k$.

The average force is determined by the momentum imparted during the absorption process ($\hbar k$), and the rate of absorption processes, the remaining terms present in Eq. (2.9). As visible in the last term in Eq. (2.9), if $I/I_0 \gg 1$ and $I/I_0 \gg (2(\delta\omega - \mathbf{k}\cdot\mathbf{v})/\Gamma)^2$, the absorption rate reaches its maximum value as $\Gamma/2$, thereby the term saturation intensity for I_0. The maximum value of the slowing force is limited by the decay rate of the electronic transition between the two energy levels and the capability to keep the light beam in resonance with the selected atomic transition. Low light intensity, or a significant frequency detuning also due to the concomitant Doppler effect, will decrease the force opposing the atom motion, thereby the need to use high-intensity light sources as the one provided by lasers, on top of maintaining the absorption process always in resonance.

This phenomenon can be used to decelerate an atomic beam, until the majority of the atoms can be trapped, by shining a laser beam head-on with the atomic beam. In order to keep the photons always on resonance with the atomic transition, it is therefore necessary either to tune accordingly the laser frequency, with use of a chirped frequency, or to tune the energy levels using the Zeeman effect with a properly tailored external magnetic field. In either case, some laser power will be also devoted to repump the atoms on the chosen recycling transition. Devices based on these two concepts, called atomic slowers, have been demonstrated in both variants, chirped laser slower and Zeeman slower, respectively. A sketch of a Zeeman slower including the setup for the diagnostic and the measurement of the energy distribution is shown in Fig. 2.4.

An alternative to atomic slowers is the use of two separate traps, one aimed at collecting and cooling the vapor created by heating a sample of the atoms and another trap in which atoms are then collected and trapped at high density, low

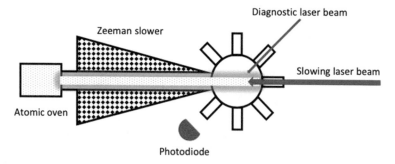

Fig. 2.4 Schematics of a Zeeman slower and the relative setup for the measurement of the atomic velocity distribution. Depending on the chosen cycling, the profile of the magnetic field can be decreasing in the direction of motion of the atomic beam, as in this case, or increasing. The latter has the disadvantage of creating large magnetic fields at the end of the slowing process, which can affect the nearby setup for subsequent atom trapping

temperature, and in a high vacuum environment, a so-called science chamber. Due to the vastly different gas pressure required in the two regions, a differential tube connecting the two traps is needed to create enough impedance to the atomic vapor. Also, the atomic sample needs to be transferred from one trap to the other, using pushing laser beams or movable magnetic traps.

2.4 Magneto-Optical Trap

The use of a single laser beam adequately tuned to the atomic transition will result in stopping the atoms and, if the process continues indefinitely, in their continuous acceleration in the direction opposite to the initial atomic motion. In order to stop the atoms in a defined location, we therefore need a different scheme. First, we need to generate an effective damping term, such that there is a viscous force with direction always opposite to the instantaneous velocity of the atom. Second, we need to create a "center of accretion" for the atoms, such that their potential energy is minimized in a well-defined location. This realizes an atomic trap in which atoms are simultaneously cooled and confined around a specific point in space [39].

The viscous force is generated by slightly detuning the laser frequency at a lower frequency with respect to the resonance corresponding to the atomic transition. The process can be described in both the laboratory frame of reference and in the frame of reference in which the atom is at rest. In the first case, the atom is moving with velocity $\mathbf{v}$, and the term $[2(\delta\omega - \mathbf{k} \cdot \mathbf{v})/\Gamma]^2$ appearing in the denominator of Eq. (2.9), considering that $\delta\omega < 0$ if the light is red-detuned from resonance, will be smaller if $\mathbf{k} \cdot \mathbf{v} < 0$, larger if $\mathbf{k} \cdot \mathbf{v} > 0$. Therefore, the absorption rate will be larger for the counterpropagating laser beam, and the recoil force will be always directed against the motion of the atom, i.e., $\mathbf{F} \cdot \mathbf{v} < 0$. In the second case, the atom will be at rest; therefore, $\mathbf{v} = 0$, but in this frame, the two laser beams will not be seen with the same frequency ω_L due to the Doppler effect. The counterpropagating beam will be seen from the atom at a frequency $\omega_L^{(-)} = \omega_L(1 + v/c)$, the beam traveling in the same direction of the atom will be seen at frequency $\omega_L^{(+)} = \omega_L(1 - v/c)$. Therefore, in this case, Eq. (2.9) still apply but with $\mathbf{v} = 0$ and the two frequencies ω_L^- and ω_L^+ for the two laser beams. Since the two frequency detunings $\delta\omega$ are different now, with $|\delta\omega^{(-)}|$ smaller than $|\delta\omega^{(+)}|$, even in this case, we have a force unbalance identical to the one evaluated in the laboratory frame, considering that $k = \omega_L/c$.

This is illustrated in part (b) of Fig. 2.3. Two counterpropagating laser beams are sent on an atom moving, in the laboratory frame, with some nonzero velocity. In the reference frame in which the atom is at rest, the two beams will be seen as red shifted (beam on the left in the figure) and blue shifted (beam on the right in the figure). If the laser is tuned at an angular frequency ω_L on the left side of the absorption resonance curve, the atom will experience different probabilities for absorption and will be preferentially excited from the counterpropagating photons. The mechanism is also working in three dimensions provided that other pairs of counterpropagating laser beams are added in a noncoplanar configuration, for instance, in the simplest case of three mutually orthogonal pairs of laser beams, as shown in Fig. 2.5. In the simplest configuration of two counterpropagating beams, the forces exerted on the atoms will be written summing two contributions as in Eq. (2.9), with opposite directions and opposite relative velocities, leading to the expression

$$\langle \mathbf{F} \rangle_{\text{cool}} = \hbar k \frac{\Gamma}{2} \frac{I}{I_0} \frac{32 k \delta\omega / \Gamma^2}{[1 + I/I_0 + (4\delta\omega^2 + k^2 v^2)/\Gamma^2]^2 - 64 k^2 v^2 \delta\omega^2 / \Gamma^4} \mathbf{v} \ . \tag{2.10}$$

in which we have regrouped the various terms as the rate of change of momentum due to the spontaneous emission rate, the relative intensity of the laser beam with respect to the saturation intensity, a form factor depending on k, I/I_0, $2\delta\omega/\Gamma$, the modulus of the atom velocity v, and finally the velocity vector $\mathbf{v}$. Equation (2.10) can be simplified by assuming an unsaturated transition ($I/I_0 \ll 1$), small velocities, and detunings such that $v \ll \delta\omega/k \ll \Gamma/k$, obtaining

$$\langle \mathbf{F} \rangle_{\text{cool}} \simeq 8 \hbar k^2 \frac{I}{I_0} \frac{2\delta\omega/\Gamma}{[1 + (2\delta\omega/\Gamma)^2]^2} \mathbf{v} \ . \tag{2.11}$$

For $\delta\omega < 0$, this creates a viscous force linear in the velocity, $\mathbf{F} = -\gamma \mathbf{v}$, with the damping coefficient γ maximized, in the simplified Eq. (2.11), if $\delta\omega = -\Gamma/(2\sqrt{3}) \simeq -0.3\Gamma$. This viscous force is directly proportional to the velocity of the atoms in three dimensions but will not allow by itself accretion of atoms around a specific location in space. The Zeeman effect can be used to make the radiation pressure dependent on space, by shifting the atomic levels in such a way that their magnetic energy is minimized in a given location, and that atomic transitions are closer to resonance when far away from this location. The simplest configuration is to have magnetic fields linearly depending on the distance from a center. This is achieved in a magnetic quadrupole configuration, easy to obtain by using two coils with currents flowing in opposite verses. The presence of magnetic fields and Zeeman shifts removing the degeneracy in m_F implies the use of laser beams with defined circular polarization, to be determined in such a way that the atoms experience the same cyclic transition regardless of their position and are progressively pushed toward the center (see Fig. 2.5). The combination of Doppler cooling and spatially dependent Zeeman shifts for trapping gives rise to the so-called magneto-optical trap (MOT).

The minimum temperature achievable in a MOT is limited by the linewidth of the atomic transition, the so-called Doppler temperature $T \simeq \hbar\Gamma/k_B$, with Γ the atomic linewidth. To derive this relationship, we need to consider the energy fluctuations occurring during the laser cooling process. The atoms undergo a random walk as they absorb and emit photons, with mean-square momentum linearly increasing

2.4 Magneto-Optical Trap

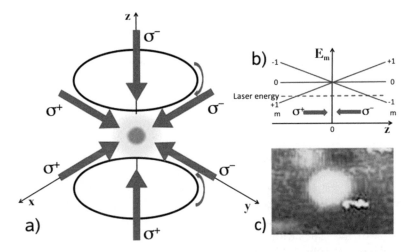

Fig. 2.5 Magneto-optical trapping. (**a**) Schematic of the counterpropagating laser beams with defined circular polarizations, quadrupole magnetic field, and Zeeman shift arrangement for one-dimensional trapping and cooling of atoms. (**b**) Energy profile for the magnetic sublevels to create a trap center with a restoring force. (**c**) Fluorescence from sodium atoms trapped in a MOT, inside a glass cell, forming an approximate sphere with 8 mm diameter. Glass cells have the advantage of a better optical access and smaller volumes with respect to stainless steel vacuum chambers, at the price of their fragility and fabrication cost

with the number of scattering events, so the square momentum changes with a rate

$$\frac{d}{dt}\langle p^2 \rangle = 2R\hbar^2 k^2 = 2D_p, \tag{2.12}$$

where we indicate with R the photon scattering rate and D_p is the diffusion coefficient of momentum. The dissipated power in the scattering process, coinciding with the cooling power, is $\mathbf{F} \cdot \mathbf{v} = -\gamma v^2$. At equilibrium, cooling and heating rates should equalize each other; therefore, $2D_p = \gamma v^2$, but at equilibrium, $mv^2 = k_B T$ due to equipartition. Considering the expression of R in terms of the laser intensity and detuning, we obtain, for low intensity and small velocity

$$k_B T = \frac{\hbar \Gamma}{4} \frac{1 + (2\delta\omega/\Gamma)^2}{2\delta\omega/\Gamma}, \tag{2.13}$$

which is minimized for $\delta\omega = \Gamma/2$, yielding the Doppler temperature

$$k_B T_D = \frac{\hbar \Gamma}{2}. \tag{2.14}$$

This limit is actually circumvented, yielding sub-Doppler cooling, an effect actually first observed as an unexpected gift and later interpreted by including the effect of the potential energy due to the laser field and including the atomic multilevel structure. In this case, the next limitation is moved to lower temperatures determined by the recoil energy of each photon absorption or emission process. This can be evaluated by considering the recoil velocity such that $mv_{\text{rec}} = \hbar k = h/\lambda$, which imply a recoil kinetic energy

$$E_{\text{rec}} = \frac{1}{2}mv_{\text{rec}}^2 = \frac{h^2}{2m\lambda^2} = k_B T_{\text{rec}}, \quad (2.15)$$

conveniently translated in terms of a corresponding recoil temperature in the last expression. Depending on the atomic species, and most importantly on their mass, the recoil temperature can be a couple of orders of magnitude smaller than the Doppler temperature. In the examples reported in Table 2.1, the recoil temperature for the lightest atom, ^{6}Li, is about 40 times smaller than the Doppler temperature, but for ^{87}Rb, the same factor increases to 403. The case of ^{133}Cs seems even more favorable, but its collisional properties have limited its convenience for achieving Bose-Einstein condensation. Moreover, the temperature for achieving quantum degeneracy also scales inversely with the mass, so in this regard a gain is only possible choosing atoms with narrow transitions, and this occurs, for instance, in the case of the various stable isotopes of ytterbium, for instance, $\Gamma = 182$ kHz for the $^1S_0 - {}^3P_1$ transition of ^{170}Yb at a wavelength of 556 nm. Further gains are possible by evading the recoil limit with so-called subrecoil cooling, but these ingenious techniques have limitations in the maximum achievable density and have not played a significant role in reaching quantum degeneracy. At high atomic density, there is excessive multiple scattering by the photons, originating reheating of the atoms. The atomic density can be increased by sheltering atoms from optical pumping in the central region of confinement, in such a way that they undergo the cycling transition only in the periphery of the cloud. This stratagem is named dark spontaneous-force optical trap or dark-SPOT. For typical atomic species, the minimum achievable temperature in typical MOT configurations is of the order of $1-10\,\mu$K, with an atomic density not exceeding 10^{12} atoms/cm^3. These values do not allow to reach the quantum degenerate regime; therefore, further cooling techniques are required, with a MOT used only as a precooling stage.

2.5 Magnetic Traps

Further cooling is achieved by other techniques in which trapping and cooling are decoupled, unlike it happens in the case of a MOT. In one class of traps, the magnetic properties of atoms are exploited, whenever present as in the case of alkali atoms. Atoms with magnetic properties, manifesting themselves as magnetic dipoles to the lowest order, will react to the presence of an external magnetic field, trying to minimize the magnetic potential energy defined as

$$U_m(\mathbf{r}) = -\boldsymbol{\mu} \cdot \mathbf{B}(\mathbf{r}). \quad (2.16)$$

2.5 Magnetic Traps

Thinking in classical terms, in a uniform magnetic field, the magnetic energy is minimized with the magnetic moment of the atom along the same direction of the magnetic field, since $U_m(\mathbf{r}) = -\mu_m B(\mathbf{r}) \cos\theta$, where μ_m and $B(\mathbf{r})$ are the moduli of the vectors $\boldsymbol{\mu}_m$ and $\mathbf{B}(\mathbf{r})$, and θ the angle between the two vectors, forming a continuum. Minimization of the magnetic energy occurs then for $\theta = 0$, maximization for $\theta = \pi$. If the magnetic field is not uniform, there will be a magnetic force expressed in terms of the gradient of the magnetic energy as

$$\mathbf{F}_m(\mathbf{r}) = -\nabla U(\mathbf{r}). \tag{2.17}$$

Using energy arguments, the atoms will move toward regions in which their magnetic (or in general any potential) energy will be minimized. For atoms in the ground state, in this case atoms with the magnetic moment aligned toward the direction of the magnetic field, they will move in the direction where the magnetic field increases (and therefore the magnetic energy becomes more negative). If the goal of the magnetic configuration is to trap atoms in a defined location, this means we need to design the magnetic field to have a maximum in a given point. It turns out that this is impossible for a static configuration, the so-called Earnshaw theorem [10], valid for both electric and magnetic fields. Therefore, we have to turn the attention to magnetic field configurations for which the magnetic field admits, in a point, a minimum value. This is possible, for instance, using two circular coils sharing the same symmetry axis and lying along two parallel planes, by running opposite currents of the same amplitude (anti-Helmholtz coils). It is left as an exercise to show that in the center of this configuration the magnetic field is zero and that the magnetic field linearly increases away from the center, with field lines outcoming from the symmetry axis and incoming along the plane orthogonal to the symmetry axis or vice versa (as a consequence of the divergenceless character of the magnetic field). The price to be paid is that the atoms must be prepared in a *low-field seeking state*, with magnetic moment antialigned to the magnetic field, a state apparently fragile with respect to the possibility of decreasing the internal energy by spin-flip. Quantum mechanics helps in this regard creating conditions for robustness of the otherwise unstable antialigned state. The full quantum mechanical expression for the magnetic energy replacing the classical expression is

$$U_m(\mathbf{r}) = g\, m_F \mu_B B(\mathbf{r}), \tag{2.18}$$

with g the gyromagnetic ratio (also known as g-factor or Landé factor) and m_F the eigenvalue of the angular momentum operator $\hat{F}$. This means that only a discrete number of magnetic energies are possible and that transitions between them cannot occur by trading infinitesimal amounts of energy, as it can happen in the classical case. These magnetic states are relatively robust as long as the rate of change of the magnetic field experienced by the atoms during their motion is much smaller than their spatially dependent Larmor frequency:

$$\frac{d\theta}{dt} \ll \omega_L = \frac{\mu_m B(\mathbf{r})}{\hbar F}. \tag{2.19}$$

Basically, atoms can be imagined as microscopic gyroscopes, and for this dynamical reasons, antitrapped states are stable in the presence of an external magnetic field, in analogy to how spinning tops are stable above a horizontal plane in the presence of vertical gravity. This stability is however violated close to a region of space in which the magnetic field is so small to make their Larmor frequency comparable or smaller to the characteristic frequency of their translational motion. This result in a high probability for spin-flip into untrapped states [29]. Even here the analogy with a spinning top holds, as the otherwise stable state above a horizontal plane becomes unstable when the spinning angular velocity becomes small and the spinning top moves below the horizontal plane. Unfortunately, the region in which the atoms are more prone to spin-flips also coincides with the accretion point of the magnetic trap, where we expect the highest density and then the most favorable conditions for the onset of quantum degeneracy. In other words, we expect, unless we take some precautions, large losses of atoms around the trapping center, just in the region where we expect to find the less energetic atoms, the premier candidates for getting quantum degeneracy. Therefore, we need to keep these atoms away from the center of the trap or to create a local minimum in the center with a nonzero value of the magnetic field, such that the inequality in Eq. (2.19) holds.

In the case of anti-Helmholtz coils, this issue is solved by using a time-orbiting potential or to plug the center with a blue-detuned laser beam.

In the first case, the zero of the magnetic field is moved away from the center of the coils, by using a time-dependent magnetic field uniformly rotating in the plane orthogonal to the symmetry axis of the quadrupole coils [37]. The rotation frequency is chosen as intermediate between the oscillation frequency of the atoms in the trap and the Larmor frequency. This makes sure that on one hand the atoms experience a potential varying so quickly that one can consider the time-average potential as the one effectively felt by the atoms, but on the other hand, it is slower than the Larmor frequency therefore inhibiting Majorana losses. With the pure quadrupole trap magnetic fields having Cartesian components

$$B_x(\mathbf{r}) = B'_x(\mathbf{r})x, \quad B_y(\mathbf{r}) = B'_y(\mathbf{r})y, \quad B_z(\mathbf{r}) = B'_z(\mathbf{r})z, \tag{2.20}$$

and since by axial symmetry $B'_x(\mathbf{r}) = B'_y(\mathbf{r})$, considering that the magnetic field must be divergenceless, we have $B'_z(\mathbf{r}) = 2B'_x(\mathbf{r})$. On top of this magnetic field component, there will be a variable magnetic field with components only in the $x - y$ plane

$$B_x^{TOP}(\mathbf{r}, t) = B_0 \cos(\omega_0 t), \quad B_y^{TOP}(\mathbf{r}, t) = B_0 \sin(\omega_0 t), \quad B_z^{TOP}(\mathbf{r}, t) = 0. \tag{2.21}$$

2.5 Magnetic Traps

It is left as an exercise to show that, once considered the total magnitude of the magnetic field, and averaged over one (or a multiple of the) period $T_0 = 2\pi/\omega_0$, the magnetic potential can be written as

$$U_{TOP}(\mathbf{r}) = \frac{\mu_m}{2}\left[B''_{rad}(x^2+y^2) + B''_z z^2\right], \qquad (2.22)$$

with the magnetic field curvatures along the radial direction $B''_{rad} = B'^2/(2B_0)$ and along the axial direction $B''_z = 4B'^2/B_0$. It is worth noticing that the magnetic potential linearly dependent on the distance from the center in a quadrupole trap is now morphed into a quadratic dependence in all directions, thereby realizing harmonic trapping, and that in the center of the trap the magnetic field is different from zero. The zero of the magnetic field will indeed occur at points in which the rotating magnetic field due to the auxiliary coils cancels the magnetic field long the plane orthogonal to the symmetry axis and passing through the center of the trap due to the quadrupole trap. This point is rotating around the center with a fixed radius equal to B_0/B'; see also sketch (a) in Fig. 2.6. Along this circumference, atoms will be lost, thereby the picturesque and dramatic term of "circle of death," as all Majorana losses are occurring along this circle. This configuration has provided the first Bose-Einstein condensates ever, of rubidium atoms, at JILA Boulder, in 1995 [2].

An alternative to the use of a time-orbiting potential is the use of a blue-detuned beam focused on the center of the trap, providing a repulsive potential. While the form of the effective potential felt by the atoms in this combination of magnetic

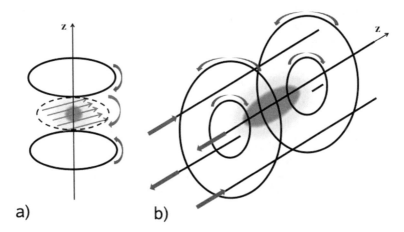

Fig. 2.6 Magnetic traps. (**a**) TOP trap with two coils in the anti-Helmholtz configuration, a rotating magnetic field and (dashed lines) the circle of death on which the magnetic field is instantaneously zero. (**b**) Ioffe-Pritchard trap, with four bars for the radial confinement, two "curvature" coils in the Helmholtz configuration for the confinement along the z-axis defining the axial direction. Coaxial with the curvature coils are two larger coils used to compensate for the bias field B_0 without significantly subtracting curvature

trap and optical antitrap is hard to modelize, this is a simple and pragmatic solution which resulted in the first Bose-Einstein condensates of sodium atoms at MIT [9].

As mentioned earlier, one can use a static configuration of magnetic fields provided that in the center of the trap the magnetic field is large enough. This is realized by using the so-called Ioffe-Pritchard trap, first used for the confinement of high-temperature plasmas [15, 38]. This configuration consists of four linear coils running the same current in opposite sense for any two neighboring coils, forming a square in a plane orthogonal to the coils. This ensures a magnetic field in the same plane whose amplitude is linearly depending on the center of the square and always directed radially. The confinement along the direction of the coils is ensured by two circular coils with axis along the symmetry axis of the four linear coils, operated in the Helmholtz configuration; see sketch (b) in Fig. 2.6. This creates a nonzero magnetic field in the midpoint of the two coils, therefore avoiding the Majorana spin-flips, and a retaining magnetic force along the symmetry axis close to the midpoint. The trapping configuration is very sensitive to the value of the magnetic field in the center, and for fine-tuning and control of the trapping frequencies, a pair of larger coils canceling most of the magnetic field without adding significant curvature is usually added; see Fig. 2.6.

The magnetic field components in a Ioffe-Pritchard trap are written as

$$B_x = B'x + \frac{B''}{2}xz, \quad B_y = B'y - \frac{B''}{2}yz, \quad B_z = B_0 + \frac{B''}{2}\left[z^2 - \frac{1}{2}(x^2 + y^2)\right], \quad (2.23)$$

where B' is the gradient created by the four linear coils in the radial direction (along the $x - y$ plane) and B_0, B'' are the magnetic field in the center of the trap ("bias field") and the curvature contribution due to the Helmholtz coils, respectively. For atoms at a temperature $k_B T < \mu_m B_0$, the trapping potential can be approximated as

$$U \simeq \frac{\mu_m}{2}\left(B''_\rho \rho^2 + B'' z^2\right), \quad (2.24)$$

where the radial curvature being $B''_\rho = B'^2/B_0 - B''/2$. This shows the sensitivity of the radial trapping B_0, thereby the need for a careful tuning of its value, and the request for stable electric currents in the circular coils. This configuration, at the price of a significantly more elaborate design for the coils, has the advantage, with respect to the quadrupole traps, of generating trapping potentials of flexible aspect ratios.

2.6 Optical Dipole Traps

Atoms are also endowed with electric properties. Due to their neutral character, they interact at the lowest order via their electric dipoles, and their electric dipole energy will be written as

$$U_e(\mathbf{r}) = -\mathbf{p} \cdot \mathbf{E}(\mathbf{r}), \tag{2.25}$$

where in turn the electric dipole moment $\mathbf{p}$ is, at least in a linear approximation, directly proportional to the electric field via the electric polarizability α, $\mathbf{p} = \alpha \mathbf{E}$ in the simplest case of isotropic electric polarizability. As in the case of magnetic fields, atoms will move toward regions for which the electric dipole energy is minimized, giving rise to the possibility of trapping in the so-called optical dipole traps [5]. A major difference from the magnetic traps is that there is no counterpart of the "gyroscopic" stabilization present for magnetic moments antiparallel to the magnetic field. However, since the electric polarizability is a dispersive phenomenon, with $\alpha = \alpha(\omega)$ depending on the frequency with which the electric field is varied [32], it is possible to distinguish two cases for the response of an atomic electric dipole to a time-dependent electric field.

In the case of static and uniform electric fields, the electric dipole in its ground state will be oriented parallel to the field lines of the electric field. In the presence of an electric field gradient, the atoms will move toward the regions in which the electric field is maximum. However, this static case, for the goal of trapping atoms around a point, is ruled out for the same reason as for magnetic fields, the Earnshaw theorem. This no-go theorem suggests the use of electric fields varying slowly with respect to the characteristic timescales of the atoms, coinciding with the reciprocal of the dominant atomic transition frequencies. In this case, the electric dipoles are still fast enough to follow the electric field direction, and a local maximum of the electric field can be created by increasing the electromagnetic energy around a point, for instance, by focusing electromagnetic waves.

If the electric field varies too quickly, the electric dipole cannot follow, rather will be out of phase, a π phase shift with respect to the electric field, as it also happens in a mechanical analogy if an harmonic oscillator is driven at a frequency higher than its resonance frequency. This occurs if the electric field is varied at a frequency higher than the resonant frequency of the dipole, determined by the dominant atomic transition frequency. In such a situation, the electric dipole energy will be minimized in a minimum of the electric field.

Therefore, the atoms will move toward regions for which the electric field is maximized if the electric field varies with a frequency lower than the atomic transition frequencies, the opposite will happen for atoms in electric fields varying with a frequency higher than the atomic transition frequencies. This second case does not allow for confining the atoms around a specific point, but it may be useful to confine atoms in a cube, a square, or a line, for instance, by using laser sheets (obtained by focusing with cylindrical lenses) along three, two, or one directions, respectively. Bosons have been condensed using blue-detuned laser beams just by compressing the cloud in a smaller volume, realizing Bose-Einstein condensation at constant temperature and increasing density. Blue-detuned beams are also crucial wherever there is the need to avoid the presence of atoms in a given region of space, like to plug the Majorana loss hole in the center of a magnetic trap, or in stirring atoms with minimal heating.

One of the distinctive advantages of optical traps with respect to magnetic traps is the accessibility of all possible hyperfine states, as the trapping mechanism does not discriminate on the basis of the magnetic moment. This allows the study of "spinor" Bose condensates, in which mixtures of atoms with different orientation of the magnetic moment are simultaneously present in the trap. Also, they leave freedom for using a magnetic field of arbitrary intensity, and the spatial and temporal manipulation of the atomic clouds occurs on micrometer and microsecond scale, respectively, features precluded to typical magnetic traps. In earlier demonstrations of these traps, the atoms were initially cooled in a MOT and a subsequent sub-Doppler stage followed by evaporative cooling in a magnetic trap [40], while later on, the laser power reached the level to trap atoms directly from a MOT.

It is not convenient to drive the atoms with electromagnetic waves having frequency too close or coinciding with the dominant transition frequencies, as there will be overwhelming absorption and emission of photons, Rayleigh scattering, which will heat the cloud due to atom recoil. In practice, electric fields from lasers are utilized, both at low frequency, so-called "red detuned" lasers as they operate at wavelengths larger than atomic transitions, and high frequency, so-called "blue-detuned" lasers, for repelling atoms. The first class will trap the atoms in the point of maximum intensity; therefore, a simple focusing of the beam in a point will create a trap provided that the laser intensity provides enough trap depth for a given initial temperature of the atomic cloud. Estimates of the trap depth are available using formulas of Gaussian optics, with the beam characterized by a nonzero minimum width at the focal point, the waist.

More quantitatively, the effective potential energy in an optical trap is given by

$$U(\mathbf{r}) = -\frac{\hbar \Omega_R^2(\mathbf{r})}{4} \left(\frac{1}{\omega_0 - \omega_L} + \frac{1}{\omega_0 + \omega_L} \right), \qquad (2.26)$$

with ω_0 the angular frequency of the atomic transition, ω_L the laser frequency, $\Omega_R(\mathbf{r})$ the position-dependent Rabi frequency (see also Chap. 5, Sect. 5.2) $\Omega_R = (I(\mathbf{r})/(2I_0))^{1/2}\Gamma$, with Γ and I_0 the linewidth and the saturation intensity of the transition, respectively. The laser intensity $I(\mathbf{r})$ depends on the selected spatial mode and for a Gaussian mode of a laser having power P will be expressed in cylindrical coordinates (ρ, z), chosen with the symmetry axis along the direction of propagation of the laser beam

$$I(\rho, z) = \frac{2P}{\pi} \frac{e^{-2\rho^2/[w_0^2(1+(z/z_R)^2)]}}{w_0^2[1 + (z/z_R)^2]}, \qquad (2.27)$$

2.6 Optical Dipole Traps

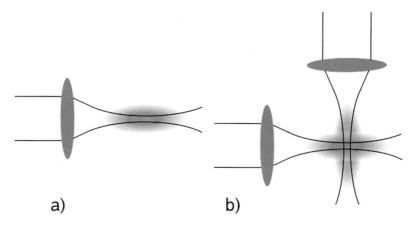

Fig. 2.7 Schematics of single beam (**a**) and crossed beam (**b**) optical dipole traps

with z_R the Rayleigh range, related to the $1/e^2$ beam waist w_0 by $z_R = \pi w_0^2/\lambda$. The spatial profile for the optical potential in this case is also Gaussian and then approximable as harmonic around the minimum of the potential. Since typically $w_0 \gg \lambda$, the axial confinement is far weaker than the radial confinement, which can be an issue in some experiments. To make stiffer the overall confinement, it is possible to cross two laser beams focused on the same point. In the simplest case of two orthogonal beams, this allows to have an almost isotropic trapping, with trapping frequencies larger than the ones usually achieved in magnetic traps. Examples of in situ images of atoms in a single beam optical dipole trap and a crossed optical dipole trap are shown in Fig. 2.7.

In addition to trapping, the effect of the laser beams is to heat the atoms via Rayleigh scattering, with a rate

$$\gamma = \left(\frac{\omega_L}{2\omega_0}\right)^3 \left(\frac{1}{\omega_0 - \omega_L} + \frac{1}{\omega_0 + \omega_L}\right)^2 \frac{I}{I_0}\Gamma^3. \tag{2.28}$$

On top of this, there can be heating due to laser noise, especially power fluctuations and beam jitter. The choice of the laser frequency is therefore obtained as a compromise between the need to maximize trap depth, requiring ω_L close to ω_0, and minimize heating, requiring the opposite. In general, it is better to use high-power, strongly detuned, lasers, due to the different power scaling with the laser frequency present in Eqs. (2.26) and (2.28). This is achieved in far-off-resonance dipole-atom trap (FORT), for instance, operating with Nd-YAG lasers around 1 μm wavelength, and so-called quasi-electrostatic traps (QUEST) carbon dioxide (CO_2) lasers, working at the even longer wavelength of 10.2 μm. The latter traps have negligible heating rates and large potential depths allowing for the formation of denegerate Bose gases [4] and Fermi gases [16], without the intermediate stage of a magnetic trap. The price to pay for the use of these traps with less heating is the need for higher laser power, in the 10 W and 100 W range for the FORT and QUEST, respectively.

Leaving aside losses of atoms via Rayleigh scattering, these optical traps are conservative; they do not have an intrinsic mechanism to cool the atoms. The possibility to decrease the trap depth by progressively decreasing the laser intensity provides a mechanism to implement evaporative cooling techniques.

We also mention the use of focused laser beams to impart forces on the atoms in specific locations and temporal patterns, with the optically plugged trap as a first example already discussed. By using programmable acousto-optical modulators, it is possible to create dimples and bumps for the atoms of arbitrary shapes, using red-detuned and blue-detuned laser beams, respectively. If the modulation frequency is large enough, this will create nearly conservative trapping with arbitrary patterns, as shown in the plots of the top left part of Fig. 2.8. In this case, the focused laser

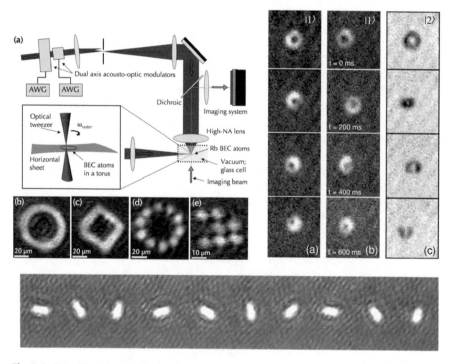

Fig. 2.8 Bose-Einstein condensates under the action of various optical potentials. (Upper left) a laser beam is focused on a sheet of light (**a**) and moved to create condensate patterns of a torus (**b**), a square (**c**), a lattice ring (**d**), and a square lattice with two defects (**e**). Credit: reprinted figure with permission from [19]. Copyright 2009 by the Deutsche Physikalische Gesellschaft. Reproduced by permission of IOP Publishing. (Upper right) vortices generated by phase imprinting using a rotating laser, with four in situ images of two hyperfine states. Credit: reprinted figure with permission from [30]. Copyright 1997 by the American Physical Society. (Bottom) Condensate put in rotation by red-detuned laser beams, 10 in situ images taken every 2 ms. Credit: reprinted figure with permission from [35]. Copyright 1999 by the American Physical Society

can be rotated in the horizontal plane by means of two acousto-optical deflectors driven by arbitrary waveform generators (AWGs) to generate any desired pattern for the time-averaged optical potential, as shown in four relevant examples, most notably a condensate in a toroidal trap. Moreover, this technique allows also to create dynamical, time-dependent patterns, which can be used, for instance, to imprint linear or angular momentum to the atoms, a feature which has been crucial in studies of the transport properties of quantum matter. Lasers have been used to imprint nonuniform phases to the condensate and then to imprint a velocity field (see top-right of Fig. 2.8 for the resulting evolution in form of quantized vortices) and to mechanically rotate a condensate to study surface waves; see bottom image of Fig. 2.8. Laser sheets have been also used to confine the atomic cloud within repulsive walls, thereby realizing even the prototypical case of infinite cubic wells always described in textbooks as first example of Bose-Einstein condensation [13].

Finally, optical dipole traps offer the possibility to work with specific wavelengths of the laser field equalizing the resulting electric polarizability for two relevant electronic states, the "magic" wavelengths [22, 50]. This allows to prepare atomic states insensitive to ac-Stark shifts, which is of relevance, for instance, in atomic clocks based on neutral atoms trapped in optical dipole potentials.

2.7 Optical Lattices

A variant of optical dipole traps is obtained by retroreflecting the laser beam to obtain a standing wave. Under this condition, the atoms experience a stationary potential with maxima and minima, of sinusoidal shape, with spatial periodicity equal to half the wavelength of the laser beam, due to the quadratic relationship between the electric field of the wave and its energy. If the setup is replicated with other two standing waves in different planes, this realizes a three-dimensional potential with spatial periodicity, an optical lattice. This allows to study quantum systems with spatial periodicity, the analogous of crystals in condensed matter systems. In the simplest three-dimensional configuration, the time-averaged potential experienced by the atoms is written as

$$U(\mathbf{r}) = U_x \sin^2(kx) + U_y \sin^2(ky) + U_z \sin^2(kz), \tag{2.29}$$

where U_i ($i = x, y, z$) is related to the amplitude of the electric field E_{0i} along the direction i as $U_i = \alpha E_{0i}^2$, α is the atomic polarizability, and k is the wavevector $k = 2\pi/\lambda$. The magnitude of the electric field in the various directions can be chosen to create lower effective dimensionality for the atoms, as depicted in Fig. 2.9, by increasing the potential energy in the trap to the point that it overwhelms the interatomic energy and, at nonzero temperature, the thermal energy. In general, the large curvature achievable in optical lattices allows for high trapping frequencies, and the lengthscale of the potential, order of the optical wavelength, allows for hopping between different sites through thermal activation or quantum tunneling.

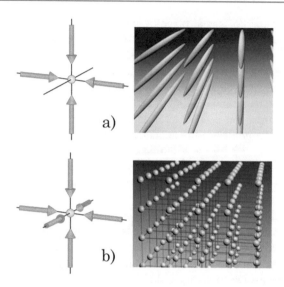

Fig. 2.9 Schematics of trapped atoms in optical lattices. If three sets of laser beams are used, the atoms will experience tight confinement around the minimum of the optical potential in all directions (**b**). If two sets of beams are used, the confinement in one direction is much weaker and only due to the possible presence of magnetic trapping or a single-beam optical dipole trapping (**a**). In this case, atoms experience a tight confinement in two dimensions only, with the possibility to behave as an effective one-dimensional gas, depending on the comparison between their temperature and the quanta of the trapping frequencies along the three axes. Credit: reprinted figure with permission from [6]. Copyright 2005 by Springer Nature

The advantages of these "human-made" lattices, also considered as examples of "synthetic" matter, are numerous. The spatial periodicity and the depth of the potential are determined by wavelength and power of the laser beams, and with high laser powers, strongly correlated systems are available. The potential varies on the wavelength scale; therefore, it is possible to set up parameters to study quantum tunneling, or thermal hopping, of atoms between adjacent sites. Impurities in the crystal can be introduced by addressing individual lattice sites with auxiliary laser beams, and lattice disorder can also be emulated by superimposing an optical speckle field. The geometry of the lattice can be chosen based on the direction of the three laser beams. Most importantly, with the lattice step determined by the laser wavelength, typically in the visible or near infrared, the individual sites of the lattice become accessible with optical microscopes, allowing for monitoring the dynamics at a level precluded in the case of usual crystals living at the nanometer scale. The modulation of the laser power allows to study the dynamics of phase transitions, also driven by control parameters other than temperature, the so-called quantum phase transitions, as we will describe in Chap. 4.

All this flexibility allows for the quantitative study of many model Hamiltonians conjectured to play a role in condensed state systems. In optical lattices, these model Hamiltonians can be realized singularly, without being masked by other effects.

In this sense, it is common to indicate these realizations as "emulating" the real system. These emulators are obviously not a replacement of the actual condensed matter system but allow to pinpoint a specific effect and to study it with accuracy only limited by the experimental precision in controlling and imaging; they can be therefore considered as analog computers at the micrometer scale.

2.8 Evaporative and Sympathetic Cooling

The two classes of traps described in the previous two sections are intrinsically conservative, as they rely on magnetic and electric dipole moments of the atoms, respectively, related to magnetic and electric potential energies. As such, the total energy is in principle conserved and no cooling may occur, while in practice, there is continuous heating due, for instance, to the background residual gas in the trap. Adiabatic techniques consisting in slowly expanding the atomic cloud by decreasing the trapping frequencies are effective in lowering the temperature of the gas but at the price of decreasing accordingly the density, such that the phase-space density is, at the best, the initial one, then with no progress toward reaching quantum degeneracy.

Atomic clouds may be instead cooled down by selectively removing atoms, the so-called evaporative cooling, a technique that is intrinsically based on many-body properties of the gas. The idea is based on the fact that temperature is related to the average energy of the atoms but that atoms follow a continuous distribution of energy, for instance, a Maxwell-Boltzmann distribution in the classical limit. Therefore, the gas will have atoms at both higher and lower energy with respect to the average value. If a mechanism is designed to remove the atoms having more energy than their average share, then the remaining atoms could, after a rethermalization stage, will settle in a new Maxwell-Boltzmann distribution at a lower temperature. A crucial point is the choice of the threshold energy above which the atoms are removed from the trap. Ideally, a large threshold energy removes significant amount of energy, but the probability that some atoms have such an excess of energy is low; therefore, it takes a long time to successfully evaporate them. Vice versa, using a low threshold energy will result in a speed up of the removal of atoms, however resulting in modest amounts of removed energy. As usual, a compromise between energy threshold and cooling speed, based on the lifetime of the atoms in the trap and all concomitant sources of inelastic collisions, yields an optimized cooling rate. Another important point is that, in order to maintain the cooling rate roughly constant while the temperature decreases, and consequently also the collisional rate decreases, the threshold energy must be lowered as cooling proceeds, for instance, in such a way that its ratio with the average thermal energy is kept constant $\eta = E_{thr}/(k_B T) = $ const, with E_{thr} indicating the threshold energy. In this condition, we speak of forced evaporative cooling, the more immediate everyday example being the act of blowing over the vapor coming out from a hot cup of coffee or tea, instead of just waiting for its natural occurrence (see Fig. 2.10).

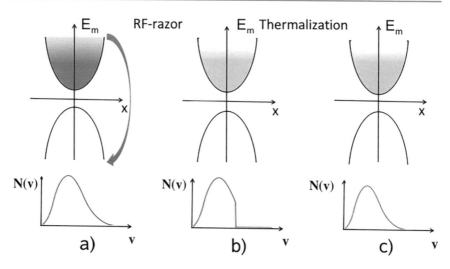

Fig. 2.10 Forced evaporative cooling in a magnetic trap. (**a**) A radiofrequency signal tuned at the difference between the magnetic energy for trapped and antitrapped atoms at a given distance from the center of the trap is applied, resulting in spin-flip and loss of the more energetic atoms which can afford to move in the peripheric region of the trap. (**b**) The intermediate state of the remaining atoms is highly nonthermal, as shown by the truncated Maxwell-Boltzmann velocity distribution. (**c**) After some time, depending on the atom density and the elastic scattering length, the remaining atoms are back in a Boltzmann state, with lower temperature. The removal of the atoms may be also achieved by lowering the trap depth, as in cooling of hydrogen or in optical dipole traps

In a harmonic potential, with the example reported in [23] choosing $\eta = 6$, 0.7 % of atoms are removed each collision time. This results in an increase of the phase space density by a factor 10^6 in 600 collision times, at the price of dealing with 1% of the initial atoms. Such a factor in phase space is in the ballpark of the one required from a precooling stage of the MOTs to achieve Bose-Einstein condensation, showing that evaporative cooling is bridging the gap between optical cooling techniques and the requirements for quantum degeneracy.

A relevant figure of merit for an evaporative cooling strategy is the dimensionless parameter α defined in terms of the time derivatives for the temperature, $\dot{T}$, and the number of trapped atoms, $\dot{N}$, as [23]

$$\alpha = \frac{\dot{T}/T}{\dot{N}/N}, \qquad (2.30)$$

where N is the number of atoms and T their temperature. Efficient evaporative cooling is achieved with large values of this parameter, as α indicates the drop in temperature per unit of atom removed from the trap. Typically, α is order of unity; therefore, a decrease in temperature by a factor 10 is accompanied by a tenfold loss of atoms. For the goal of achieving the deepest quantum degeneracy, sheer cooling is not enough; therefore, one has to also introduce another figure of merit as $\rho =$

$-(\dot{D}/D)/(\dot{N}/N)$, where D is the phase space density introduced earlier. Progress toward quantum degeneracy can be stalled if α is constant or increases while instead ρ is constant or decreases, as it could happen for an adiabatic expansion.

This cooling technique is intrinsically many body, therefore does not suffer from the limitations of single-atom optical cooling in which high densities of the cloud produce light rescattering and atom recoil. As such, it is efficient over a broad range of densities and temperatures, with only practical limitations specific to each apparatus and experiment assigned some peculiarities of the atomic species in exam. The crucial factor is to ensure the dominance of elastic collisions over all other sources of collisions in which the internal state of the atoms is changed, such as dipolar relaxation or three-body collisions with formation of molecules, and of course background gases which implies the use of ultra-high vacuum techniques, with ion and titanium sublimation pumps, initially backing out the entire apparatus for several days. The mix of precautions to achieve efficient evaporative cooling depends on the atomic species and on the density of the cloud. For this reason, evaporative cooling is usually accompanied, whenever possible, by an increase in the atomic density stiffening the trapping potential. A drawback of evaporative cooling is the diminished number of atoms remaining in the trap, which make compulsory to start with the larger number of atoms trapped in the MOT precooling stage. Also, some species, such as ^{133}Cs, are recalcitrant to rethermalization due to a low elastic scattering length, although this can be circumvented using external magnetic fields tuned close to Feshbach resonances.

The concrete techniques to selectively remove atoms differ for magnetic and optical traps.

In the first case, one relies on the magnetic energy of the atoms and the fact that trapped atoms are oriented with their magnetic dipoles antiparallel to the local magnetic field. As discussed earlier, this configuration is unstable from the classical standpoint, and its stability is only enforced in a quantum mechanical setting due to the conservation of m_F. Such a stability can be upset if a photon is absorbed with frequency equal to the difference between the magnetic energy with the magnetic moment oriented antiparallel to the local magnetic field and the magnetic energy with the magnetic moment parallel, which is $\Delta E_{\text{hf}} = 2\mu B(\mathbf{r})$. Then, a spin-flip will occur with high probability and the resulting atom will feel a magnetic force pointing toward a region in which the amplitude of the magnetic field is larger. This realizes a condition of antitrapping, and the atoms will be quickly removed from the trap. The magnetic energy difference depends on the location of the atoms and is larger in the outer region of the trap, where more energetic atoms are orbiting around the center. Thus, one can devise a forced cooling strategy in which a chirp of radiofrequency is sent to the atoms with progressively lower frequency, until it stops short of the frequency corresponding to the magnetic field very close to the center of the trap. Pictorially, this is referred to as "shaving" the atomic cloud. Care should be exerted in compensating the lower number of atoms, thereby the lower density, by stiffening the trapping potential in such a way that rethermalization always occurs in a timescale much shorter than the lifetime of the atoms in the trap. Keeping in mind that $h/k_B = 48$ μK/MHz, the photon energy falls in the radiofrequency range; therefore, a coil used as an antenna coupled to a rf-generator with programmable frequency and amplitude is all one needs to implement rf-forced evaporative cooling.

Radiofrequencies can also be used for other tasks. In particular, they provide a shield for low density atoms still trapped but at quite large distance from the center and with energy of order of mK. In spite of their low density and scarce overlap with the colder component near the center of the trap, occasional encounters may cause substantial heating due to their higher temperature. This cloud of atoms orbiting at large distance is also referred to as "Oort cloud," in analogy with the conjectured halo of comets at large distance from the planetary system but still within the gravitational attraction of the sun. A radiofrequency tuned at a value corresponding to a large surface will induce spin-flip of the atoms crossing the surface and in doing so will protect the ultracold atoms in the center of the trap from these occasional collision. Evidence for the Oort cloud is indirect due to the low density of the halo, preventing their imaging, but the application of the "rf-shield" significantly improves the lifetime of the atoms [7, 31]. Radiofrequencies can also be used to tailor the magnetic potential, as discussed in [12, 51], to perform spectroscopy of hyperfine states, and as an output coupler for atoms lasers, as we will describe in Chap. 5.

In the case of optical traps, evaporative cooling is achieved by simply lowering the laser power, thereby reducing the trap depth. In comparison to magnetic traps, here evaporative cooling is a more delicate issue as by lowering the laser intensity the trapping becomes weaker; therefore, the atomic density is also lower. This requires a careful optimization as lower densities imply longer timescales for the equilibration of the atomic cloud. A bonus available consists in increasing the interatomic interactions via increase of the elastic scattering length using Feshbach resonances. Since these require a uniform magnetic field, this option is not available for magnetic traps, unless optical Feshbach resonances are exploited. A similar technique, lowering the trap depth, has been used to evaporatively cool atomic hydrogen to quantum degeneracy using magnetic traps [20]. Evaporative cooling in this case has been implemented with a modulation of the escape barrier of the magnetic trap, by changing the position of the saddle potential in the axial trapping direction. The cooling efficiency and speed can decrease with evaporation occurring in lower dimensionality, depending on the rate of ergodic mixing into the other dimensions.

In general, evaporative cooling is quite effective in lowering the temperature by about one or two orders of magnitude, at the price of reducing the number of atoms by a factor of 10^3–10^4. The cooling rate depends on the lifetime of the atoms in the trap, in turn depending on the details of the pumping system and the outgassing of the vacuum components. It is necessary to use ultra-high vacuum (UHV) techniques, with ion and sublimation pumps, bakeout of the apparatus, and depending on this, the lifetime of the atoms can vary between a few seconds up to 10^2 seconds. Using a cryogenic apparatus at liquid He temperature, therefore with a very low background gas pressure of 4×10^{-12} Torr, it has been possible to achieve a lifetime of over

2.8 Evaporative and Sympathetic Cooling

one hour in a MOT and almost 10 minutes in a magnetic trap [48]. The minimum achievable temperature to cooling is therefore of practical nature and also related to the minimum number of atoms remaining in the trap still enough to be manipulated and detected for a targeted experiment.

Evaporative cooling has also peculiar limitations for Fermi gases in magnetic traps. At the lowest temperatures, elastic scattering between atoms occurs predominantly in s-wave; therefore, the angular part of the two-body wave function is symmetric. Since in a magnetic trap a well-defined spin state is trapped, the two-body wave function in spin variables is also symmetric. The need to have a globally antisymmetric wave function for identical fermions inhibits s-wave scattering, and therefore, thermalization processes are suppressed, the so-called Pauli-blocking. A possibility to overcome this limitation is to create a distinct spin state and to proceed with each atom in a defined spin state to collide atoms in the other spin state, so-called dual evaporative cooling. This implies smaller number of atoms at the end of the evaporative process, an issue particularly sensitive for fermions, significantly decreasing the signal when imaging the atomic cloud.

An alternative which allows to conserve the number of fermions during the cooling process is to create a Fermi-Bose mixture in which a Bose gas already present with large number of atoms (to maximize its heat capacity) is progressively cooled via evaporative cooling. This sympathetic cooling was first demonstrated for Bose atomic mixtures, in particular to cool ^{85}Rb through thermal contact with ^{87}Rb [33]. Provided the interspecies elastic scattering length is large enough, the Fermi gas will "follow" the Bose gas and will be progressively cooled, at least until the Bose gas will reduce its heat capacity to become comparable to the one of the Fermi gas. The need to enhance the interspecies elastic scattering makes more practical the use of optical dipole traps, as in this case a uniform magnetic field can be applied to exploit Feshbach resonances. As first evidenced by the group at Rice University, there is a limit to sympathetic cooling since the heat capacity of the Bose gas is progressively decreasing both because the number of atoms is decreasing while undergoing evaporation and because below the critical temperature for Bose-Einstein condensation the specific heat of the Bose gas decreases dramatically with the cubic power of the temperature [46]. The specific heat of the Fermi gas instead decreases only linearly with the temperature; therefore, eventually the heat capacity of the Bose gas is comparable or smaller than the one of the Fermi gas, and sympathetic cooling becomes inefficient. This issue of heat capacity matching between the Bose and the Fermi gases could be circumvented by using a bichromatic optical dipole trap, in which the trapping frequencies of the two gases can be chosen differently, in such a way that the Bose gas is maintained above the critical temperature for BEC, while the Fermi gas is still in the deep quantum degenerate regime (i.e., $\omega_F \gg \omega_B$) [36]. Bichromatic optical dipole traps have been demonstrated in three laboratories so far [43, 47, 49], but optimized sympathetic cooling by heat capacity matching has not yet been achieved, and the minimum reachable Fermi degeneracy parameter is still in the $T/T_F \simeq 0.05$ range [17].

It should also be noticed that fermions have a peculiar heating mechanism, called Fermi hole heating, which could limit the minimum reachable T/T_F even with future progress suppressing more technical sources of heating or optimized heat capacity matching with a Bose species. By referring to Fig. 1.1, if one of the fermions in the intermediate states is scattered by the atoms of the residual background pressure, a cascade of transitions to fill the resulting hole will occur, reminiscent of the Auger effect for atomic electrons. As the heat capacity of a Fermi gas tends to zero at low temperatures, this heating significantly increases the temperature of the gas, frustrating any effort to decrease T/T_F [45]. Even in this case immersion of the ultracold fermions into a large heat capacity reservoir [8] is beneficial, as demonstrated with precision calorimetry in a toroidal trap with ^{6}Li atoms in the degenerate regime [1].

A variety of techniques have been proposed and demonstrated for cooling of atoms in optical lattices [28]. The transfer of atoms from a single trap to an optical lattice in general results in heating; therefore, a further stage of cooling for the atoms while in the optical lattice is often deemed necessary. These techniques can be divided into two classes, reminiscent of the division of many-body cooling techniques for single traps as evaporative cooling and sympathetic cooling. In the first class, one filters out atoms present in modes of the system in which entropy is mainly concentrated (filter cooling), while in the second, the targeted system is immersed into a reservoir which is used as an entropy sink (immersion cooling). Even in this case, species-dependent optical lattices have been demonstrated.

2.9 Problems

2.1. Consider a thermal beam of sodium atoms at a temperature $T = 800$ K. Calculate the maximum deceleration expected if a counterpropagating laser beam resonant with the D_2 transition has a power saturating the transition and the minimum time and space to bring the atomic beam at rest, assuming that the Doppler shift is properly compensated by an external magnetic field. Estimate the transversal spread of the velocity after they stop in the longitudinal direction and the associated temperature.

2.2. The optimization of the cooling efficiency in Eq. (2.11) occurs for a $\delta\omega$ comparable to Γ (at variance with the approximation used to derive the equation, $\delta\omega \ll \Gamma$), and the resulting damping coefficient γ does not depend upon the velocity. Study the viscous force in the complete expression for the force in Eq. (2.10), and discuss both the optimization upon $\delta\omega$ and the velocity dependence of γ.

2.3. Derive the expression in Eq. (2.22), including B''_{rad} and B''_z in terms of B_0 and B'.

2.4. Derive the expression for the potential depth of an Ioffe-Pritchard trap. Discuss advantages and disadvantages in using the modulation of this potential depth for evaporative cooling with respect to radiofrequency induced evaporative cooling.

2.5. The laser power for an optical dipole trap is linearly decreased in a time duration T_0 from its initial value P_{in} to its final value P_{fin}. Initially, the trap contains a gas of nondegenerate N atoms at temperature T_{in}. Assume that atoms with an energy E such that $0.95 V_d < E < V_d$ are irreversibly lost from the trap, with V_d the trap energy depth. Under the hypothesis of a fast thermalization time τ, i.e., $\tau \ll T_0$, estimate the time dependence of the temperature of the atomic gas, if it remains nondegenerate during the entire process.

2.6. A Fermi gas of $N = 10^6$ atoms is put in contact with a Bose gas of $N = 10^8$ atoms initially at a temperature $T = 10^2 \, T_{\text{BEC}}$. Both gases share the same harmonic trapping potential, and assume very fast interspecies and intraspecies thermalization times. The Bose gas thereafter is evaporatively cooled with removal of atoms at the rate of $\dot{N} = 10^5$ atoms/s, with an evaporative coefficient $\alpha = 0.1$. Describe the time dependence of the temperature of the mixture until the Bose gas is exhausted.

2.10 Further Reading

- V. S. Letokhov, *Nonlinear high resolution spectroscopy*, Science **190**, 344 (1975).
- V. I. Balykin, V. S. Letokhov, and V. G. Minogin, *Cooling atoms by means of laser radiation pressure*, Sov. Phys. Usp. **28**, 803 (1985).
- S. Stenholm, *The semiclassical theory of laser cooling*, Rev. Mod. Phys. **58**, 699 (1986).
- C. Cohen-Tannoudji, J. Dupont-Roc, and G, Grynberg, *Atom-Photon Interactions: Basic Processes and Applications*, (Wiley, New York, 1992).
- W. Ketterle and N. V. Druten, *Evaporative Cooling of Trapped Atoms*, Advances in Atomic, Molecular, and Optical Physics, **37**, (Academic Press, New York, 1996), pp. 181–236.
- S. Chu, *Nobel Lecture: The manipulation of neutral particles*, Rev. Mod. Phys. **70**, 685 (1998).
- C. N. Cohen-Tannoudji, *Nobel Lecture: Manipulating atoms with photons*, Rev. Mod. Phys. **70**, 707 (1998).
- W. D. Phillips, *Nobel Lecture: Laser cooling and trapping of neutral atoms*, Rev. Mod. Phys. **70**, 721 (1998).
- M. Inguscio, S. Stringari, and C. E. Wieman (editors), *Bose-Einstein Condensation in Atomic Gases*, Proceedings of the International School of Physics "Enrico Fermi", Course CXL (IOS Press, Amsterdam, 1999).
- I. Bloch, *Ultracold quantum gases in optical lattices*, Nature Physics **1**, 23 (2005).
- W. Ketterle and M. Zwierlein, *Making, probing and understanding ultracold Fermi gases*, Nuovo Cimento **31**, 247 (2008).
- J. Mitroy, M. S. Safronova, and C. W. Clark, *Theory and applications of atomic and ionic polarizabilities*, J. Phys. B **43**, 202001 (2010).

References

1. Allman, D.G., Sabharwal, P., Wright, K.C.: Mitigating heating of degenerate fermions in a ring-dimple atomic trap. Phys. Rev. A **107**, 043322 (2023)
2. Anderson, M.H., Ensher, J.R., Matthews, M.R., Wieman, C.E., Cornell, E.A.: Observation of Bose-Einstein condensation in a dilute atomic vapor. Science **269**, 198 (1995)
3. Ashkin, A.: Atomic-beam deflection by resonance-radiation pressure. Phys. Rev. Lett. **25**, 1321(1970)
4. Barrett, M.D., Sauer, J.A., Chapman, M.S.: All-optical formation of an atomic Bose-Einstein condensate. Phys. Rev. Lett. **87**, 010404 (2001)
5. Bjorkholm, J.E., Freeman, R.H., Ashkin, A., Pearson, D.B.: Observation of focusing of neutral atoms by the dipole forces of resonance-radiation pressure. Phys. Rev. Lett. **41**, 1361 (1978)
6. Bloch, I.: Ultracold quantum gases in optical lattices. Nat. Phys. **1**, 23 (2005)
7. Burt, E.A., Ghrist, R.W., Myatt, C.J., Holland, M.J., Cornell, E.A., Wieman, C.E.: Coherence, correlations, and collisions: what one learns about Bose-Einstein condensates from their decay. Phys. Rev. Lett. **79**, 337 (1997)
8. Cotè, R., Onofrio, R., Timmermans, E.: Sympathetic cooling route to Bose-Einstein condensate and Fermi-liquid mixtures. Phys. Rev. A **72**, 041605(R) (2005)
9. Davis, K.B., Mewes, M.-O., Andrews, M.R., van Druten, N.J., Durfee, D.S., Kurn, D.M., Ketterle, W.: Bose-Einstein condensation in a gas of Sodium atoms. Phys. Rev. Lett. **75**, 3969 (1995)
10. Earnshaw, S.: On the nature of the molecular forces which regulate the constitution of the luminiferous ether. Trans. Cambridge Philos. Soc **7**, 97 (Read March 18, 1839 and published in 1842)
11. Frisch, R.: Experimenteller Nachweis des Einsteinschen Strahlungsruckstosses. Z. Phys. **86**, 42 (1933)
12. Garraway, B.M., Perrin, H.: Recent developments in trapping and manipulation of atoms with adiabatic potentials. J. Phys. B: At. Mol. Opt. Phys. **49**, 172001 (2016)
13. Gaunt, A.L., Schmidutz, T.F., Gotlibovych, I., Smith, R.P., Hadzibabic, Z.: Bose-Einstein condensation of atoms in a uniform potential. Phys. Rev. Lett. **110**, 200406 (2013)
14. Gehm, M.E.: Properties of ^{6}Li, from Preparation of an Optically-Trapped Degenerate Fermi Gas of 6Li: Finding the Route to Degeneracy, PhD thesis. Duke University, Durham (2003). https://jet.physics.ncsu.edu/techdocs/pdf/PropertiesOfLi.pdf
15. Gott, Y.V., Ioffe, M.S., Tel'kovskii, V.G.: Some new results on plasma confinement in a magnetic trap. Nucl. Fusion Suppl. **3**, 1045 (1962)
16. Granade, S.R., Gehm, M.E., O'Hara, K.M., Thomas, J.E.: All-optical production of a degenerate Fermi gas. Phys. Rev. Lett. **88**, 120405 (2002)
17. Hadzibabic, Z., Gupta, S., Stan, C.A., Schunck, C.H., Zwierlein, M.W., Dieckmann, K., Ketterle, W.: Fiftyfold improvement in the number of quantum degenerate fermionic atoms. Phys. Rev. Lett. **91**, 160401 (2003)
18. Hänsch, T.W., Schawlow, A.L.: Cooling of gases by laser radiation. Optics Comm. **13**, 68 (1975)
19. Henderson, K., Ryu, C., MacCormick, C., Boshier, M.G.: Experimental demonstration of painting arbitrary and dynamic potentials for Bose–Einstein condensates. New J. Phys. **11**, 043030 (2009)
20. Hess, H.F.: Evaporative cooling of magnetically trapped and compressed spin-polarized hydrogen. Phys. Rev. B **34**, 3676 (1986)
21. Humbert, L., Baker, M., Sigle, D., van Ooijen, E.D., Haine, S.A., Davis, M.J., Heckenberg, N.R., Rubinsztein-Dunlop, H.: Time-averaged optical dipole traps for Bose-Einstein condensates. In: International Quantum Electronics Conference (IQEC) and Conference on Lasers and Electro-Optics (CLEO) Pacific Rim incorporating the Australasian Conference on Optics, Lasers and Spectroscopy and the Australian Conference on Optical Fibre Technology, pp. 1780–1872 (2011)

References

22. Katori, H., Ido, T., Kuwata-Gonokami, M.: Optimal design of dipole potentials for efficient loading of Sr atoms. J. Phys. Soc. Jpn. **68**, 2479 (1999)
23. Ketterle, W., van Druten, N.J.: Evaporative cooling of trapped atoms. In: Advances in Atomic, Molecular, and Optical Physics, vol. 37, pp. 181–236. Academic Press, New York (1996)
24. Lebedev, P.N.: Untersuchungen Über die Druckkräfte des Lichtes. Annalen der Physik (Leipzig) **6**, 433 (1901)
25. Letokhov, V.S.: Narrowing of the Doppler width in a standing light wave. Pis'ma. Zh. Eksp. Teor. Fiz. **7**, 348 (1968) [JETP Lett. **7**, 272 (1968)]
26. Letokhov, V.S., Minogin, V.G., Pavlik, B.D.: Cooling and trapping of atoms and molecules by a resonant light field. Optics Comm. **19**, 72 (1976)
27. Letokhov, V.S., Minogin, V.G.: Laser radiation pressure on free atoms. Phys. Rep. **73**, 1 (1981)
28. McKay, D.C., DeMarco, B.: Cooling in strongly correlated optical lattices: prospects and challenges. Rep. Prog. Phys. **74**, 054401 (2011)
29. Majorana, E.: Atomi orientati in campo magnetico variabile. Nuovo Cimento **9**, 43 (1932)
30. Matthews, M.R., Anderson, B.P., Haljan, P.C., Hall, D.S., Wieman, C.E., Cornell, E.A.: Vortices in a Bose-Einstein condensate. Phys. Rev. Lett. **83**, 2498 (1999)
31. Mewes, M.O., Andrews, M.R., van Druten, N.J., Kurn, D.M., Durfee, D.S., Ketterle, W.: Bose-Einstein condensation in a tighly confining DC magnetic trap. Phys. Rev. Lett. **77**, 416 (1996)
32. Mitroy, J., Safronova, M.S., Clark, C.W.: Theory and applications of atomic and ionic polarizabilities. J. Phys. B **43**, 202001 (2010)
33. Myatt, C.J., Burt, E.A., Ghrist, R.W., Cornell, E.A., Wieman, C.E.: Production of two overlapping Bose-Einstein condensates by sympathetic cooling. Phys. Rev. Lett. **78**, 586 (1997)
34. Nichols, E.F., Hull, G.F.: The pressure due to radiation. Astrophys. J. **17**, 315 (1903)
35. Onofrio, R., Durfee, D.S., Raman, C., Köhl, M., Kuklewicz, C.E., Ketterle, W.: Surface excitations of a Bose-Einstein condensate. Phys. Rev. Lett. **84**, 810 (1999)
36. Onofrio, R., Presilla, C.: Reaching Fermi degeneracy in two-species optical dipole traps. Phys. Rev. Lett. **89**, 100401 (2002)
37. Petrich, W., Anderson, M.H., Ensher, J.R., Cornell, E.A.: Stable, tightly confining magnetic trap for evaporative cooling of neutral atoms. Phys. Rev. Lett. **74**, 3352 (1995)
38. Pritchard, D.E.: Cooling neutral Atoms in a magnetic trap for precision spectroscopy. Phys. Rev. Lett. **51**, 1336 (1983)
39. Raab, E.L., Prentiss, M., Cable, A., Chu, S., Pritchard, D.E.: Trapping of neutral sodium atoms with radiation pressure. Phys. Rev. Lett. **59**, 2631 (1987)
40. Stamper-Kurn, D.M., Andrews, M.R., Chikkatur, A.P., Inouye, S., Miesner, H.-J., Stenger, J., Ketterle, W.: Optical confinement of a Bose-Einstein condensate. Phys. Rev. Lett. **80**, 2027 (1998)
41. Steck, D.A.: Rubidium 87 D Line Data (2001). http://steck.us/alkalidata
42. Steck, D.A.: Sodium D Line Data (2000). http://steck.us/alkalidata
43. Tassy, S., Nemitz, N., Baumer, F., Höhl, C., Batar, A., Görlitz, A.: Sympathetic cooling in a mixture of diamagnetic and paramagnetic atoms. J. Phys. B: At. Mol. Opt. Phys. **43**, 205309 (2010)
44. Gehm, M.E.: Properties of Potassium, from 'Feshbach resonances in ultracold mixtures of the fermionic quantum gases 6Li and 40K, PhD thesis. University of Amsterdam, Amsterdam (2009). https://www.tobiastiecke.nl/archive/PotassiumProperties.pdf
45. Timmermans, E.: Degenerate fermion gas heating by hole creation. Phys. Rev. Lett. **87**, 240403 (2001)
46. Truscott, A.G., Strecker, K.E., McAlexander, W.I., Partridge, G.B., Hulet, R.G.: Observation of Fermi pressure in a gas of trapped atoms. Science **291**, 2570 (2001)
47. Vaidya, V.D., Tiamsuphat, J., Rolston, S.L., Porto, J.V.: Degenerate Bose-Fermi mixtures of rubidium and ytterbium. Phys. Rev. A **92**, 043604 (2015)
48. Willems, P.A., Libbrecht, K.G.: Creating long-lived neutral-atom traps in a cryogenic environment. Phys. Rev. A **51**, 1403 (1995)
49. Wilson, K.E., Guttridge, A., Segal, J., Cornish, S.L.: Quantum degenerate mixtures of Cs and Yb. Phys. Rev. A **103**, 033306 (2021)

50. Ye, J., Vernooy, D.W., Kimble, H.J.: Trapping of single atoms in cavity QED. Phys. Rev. Lett. **83**, 4987 (1999)
51. Zobay, O., Garraway, B.M.: Two-dimensional atom trapping in field-induced adiabatic potentials. Phys. Rev. Lett. **86**, 1195 (2001)

Ultracold Atoms as Weakly Correlated Systems 3

In this chapter, we first describe the imaging techniques for ultracold atomic gases, and the first generation of experiments in which quantum degenerate behavior of Bose and Fermi gases has been explored. Following the historical path, this implies to first focus on Bose gases, as the first evidences of quantum degenerate behavior in Fermi gases occurred about 5 years later, mainly due to the difficulties in cooling Fermi gases discussed in the previous chapter. From a theoretical standpoint, the mean-field approach on which the Gross-Pitaevskii equation relies is sufficient for the interpretation of these experiments. We then proceed to study the response of a Bose-Einstein condensate to external forces and torques. Always in a mean-field approach, this allows to describe some features of quantum transport common to superfluidity in ^{4}He, including the presence of localized solutions with nontrivial topology in form of quantized vortices. Finally, by reducing the effective dimensionality of the Bose-Einstein condensate, we consider a simpler example of topological solutions in the form of solitons.

3.1 Imaging of Ultracold Atomic Clouds

The imaging of ultracold clouds proceeds via interaction with photons, since any contact with a material, solid-state probe at higher temperatures, due to the comparatively negligible heat capacity of the cloud, will destroy the sample without affecting significantly the dynamics of the probe. Collecting light emitted from the atoms by fluorescence is not effective, because of the small signal especially for small clouds, and certainly not viable for "dark" stages of magnetically trapped atoms. Nevertheless, fluorescence imaging is a cheap and quick way to maximize the atom number in a MOT, using a photodiode, an image, or just the eye in case of fluorescence in the visible, for instance, in the case of sodium; see (a) in Fig. 3.1.

The most effective way to image small atomic clouds is to send a uniform flux of photons in resonance or near resonance with an atomic transition, with a CCD

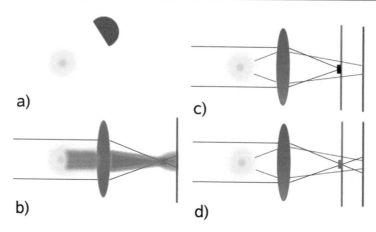

Fig. 3.1 Imaging techniques for atoms. If the atomic cloud is large enough and in the presence of light, as in a MOT, atoms can be detected by the fluorescence light, they emit by simply using a photodiode or the naked eye (**a**). Alternatively, a nearly resonant uniform beam can be sent, and one can observe the shadow of the atoms on a CCD camera (**b**). Dispersive techniques can also be used, with the advantage of a smaller heating of the cloud, as dark-ground technique, in which all unscattered light is stopped (**c**), and phase-contrast imaging, with the use of a phase plate to allow interference between the scattered and unscattered light (**d**)

camera after the atomic cloud measuring the photons surviving the interaction. The optical density, i.e., the density integrated over the thickness of the atomic cloud along the photon direction, is then obtained by using the exponential relationship between the density and the absorption coefficient. The relationship between the impinging electric field E_0 and the transmitted electric field E is

$$E = t E_0 e^{i\phi}, \qquad (3.1)$$

where t is the transmission amplitude and ϕ the phase shift induced by the cloud.

In absorption imaging, the focus is on measuring the transmission coefficient $|t|^2 = |E|^2/|E_0|^2 = I/I_0$, where I_0 is the intensity (assumed to be uniform) of the incoming light beam along the z axis and I the transmitted intensity reaching a location (x, y) of the CCD camera; see (b) in Fig. 3.1.

In terms of atomic properties, given the atomic polarizability α, the refraction index in a gas is $n_{\text{refr}} = (1 + 4\pi\alpha n)^{1/2}$, with n the atomic density. For low density gases, such that $4\pi\alpha n \ll 1$, and focusing on just one atomic transition (two-level system), we have

$$n_{\text{ref}} = 1 + \frac{n\sigma_0 \lambda}{4\pi(1+\delta)^2}(i - \delta), \qquad (3.2)$$

where σ_0 is the resonant cross section, λ is the light wavelength, and $\delta = (\omega - \omega_{at})/(\Gamma/2)$ the detuning between the light angular frequency and the atomic

3.1 Imaging of Ultracold Atomic Clouds

transition angular frequency, normalized to half the frequency linewidth, and i the imaginary unit. Light absorption is proportional to the imaginary part of the refraction index and then is maximized when in resonance, $\delta = 0$, while dispersive effects are maximized at $\delta = \pm 1$. For several atomic resonances, they will appear as added to right-hand side of Eq. (3.2), although they will weight much less if far detuned from light.

Absorption imaging is very sensitive to low density clouds, like the thermal component of the gas, but if instead the cloud is too dense or too thick, there will be a large region in which all photons will be absorbed. The absorption will then saturate making less reliable the quantitative determination of the integrated atomic density. This can easily happen for Bose-Einstein condensates at their maximum density, and it is therefore advisable, for a quantitative assessment of the density, to detune the light beam to avoid saturation. This issue is mitigated for images not taken in situ, after switching off the trapping potential and waiting a time of flight such that saturation no longer occurs. Another drawback of this imaging technique is the heating of each atom after the absorption process. The recoil energy is typically enough to expel the atoms from the trap. In general blasting, several atoms away will preclude the execution of a second imaging on the same ultracold cloud. Considering shot-to-shot fluctuations in the production of atomic samples, this may limit the quantitative study of several dynamical phenomena. In order to overcome this limitation, imaging techniques used since many decades for samples with refractive index close to unity, such as biological materials, have been used, the so-called phase-contrast imaging.

The idea is to intentionally send off-resonant light and aim at the detection of the refractive effect of the sample, converting the resulting phase shifts in intensity measurements by comparison with a reference signal. Off-resonant light will result in negligible absorption of photons; therefore, multiple images may be taken before the sample will be significantly degraded in atom number due to the cumulative residual heating [4]. The atomic cloud will induce a phase shift proportional to its refraction index, in turn depending on the cloud density. This technique has been successfully used to image in situ the formation of a Bose-Einstein condensate after a fast evaporative cooling protocol [31], among the various early applications. Two variants of phase-contrast imaging have been used, dark-ground and phase-contrast imaging. In the first, all unscattered light is stopped on a plate with a total absorption spot located on the focus of a converging lens after the sample (see (c) in Fig. 3.1). Then, only scattered light will pass through to be focused on a camera. This method is simple but has the inconvenience of a low sensitivity. If E_0 is the amplitude of the electric field of the incoming plane wave, and $E = t E_0 e^{i\phi}$ is the amplitude of the scattered light, with t the transmission amplitude and ϕ the phase shift, then stopping the unscattered light gives a signal, in intensity and time averaged, as $\langle I_{dg} \rangle \propto |E - E_0|^2$ which, in terms of the initial intensity, is

$$\langle I_{dg} \rangle = I_0(1 - 2t\cos\phi + t^2) \,. \tag{3.3}$$

For small dispersions, this yields a signal quadratic in ϕ; therefore, it is not suitable for imaging samples with small optical density. This drawback is circumvented if instead one uses as a reference the unscattered light but phase shifted by $\pm\pi/2$, (see (d) in Fig. 3.1). Operationally, this requires to replace the small absorption spot on the focus of the lens with a phase plate; in this case, the intensity collected on the camera is $\langle I_{pc} \rangle \propto |E + E_0(e^{\pm i\pi/2} - 1)|^2$ and the signal on the camera, for small ϕ, is linearly depending on ϕ.

Phase-contrast imaging has the distinctive advantage, with respect to absorption imaging, of avoiding the shot-to-shot fluctuations present in the latter, due to the limited reproducibility of atom number (and phase) in different runs of any apparatus. Its drawback is the need to have enough number of atoms and the possibility to introduce systematic effects due to the presence of imperfections in the phase plate. It is worth to remark that one of the strongest assets of ultracold atom physics in general is the possibility to study in detail the dynamics of processes. In most of condensed matter physics instead, due to the stronger interactions between the constituents, it is difficult to evidence dynamical phenomena in general as they occur too quickly. In this regard, phase-contrast imaging provides a unique technique basically limited by the density of the atomic sample and its residual heating whenever a large number of images is collected. The dispersive interaction results in phase scrambling, based on general limitations in a quantum measurement process, and then phase-contrast imaging cannot be trivially used, for instance, in repeated measurements of the intrinsic coherence of a Bose condensate.

3.2 First-Generation Experiments

Having discussed the imaging of ultracold atomic clouds, we can now proceed with the identification of quantum degenerate behavior, approximately following the historical sequence of these pioneering experiments. The first atoms to be cooled to Bose degeneracy have been ^{87}Rb [3], ^{23}Na [13], and ^{7}Li [8], while ^{40}K has been the first fermionic atom brought into quantum degenerate regime [14] and then followed by various experiments using ^{6}Li. The initial evidence for Bose-Einstein condensation in ^{7}Li was rather circumstantial and affected by spherical aberration in a lens of the imaging system, thereafter leading to a reassessment of the number of atoms in the condensed phase of about 10^3, instead of the original estimate of 2×10^5.

Bose-Einstein condensates of ^{87}Rb were first produced, at the Joint Institute for Laboratory Astrophysics (JILA) in Boulder, by a group led by Eric Cornell and Carl Wieman [3]. The atoms, arising from a vapor cell and trapped into a MOT, were transferred into a time-orbiting potential quadrupole magnetic trap. The first evidence was collected by switching off the trap and imaging the cloud while in free expansion. By changing the final frequency of the evaporative radiofrequency sweep, they observed the morphing from an optical density compatible with a Gaussian distribution of atoms from the trap center to a bimodal distribution in which, on top of the Gaussian, thermal component, it appeared a higher density

3.2 First-Generation Experiments

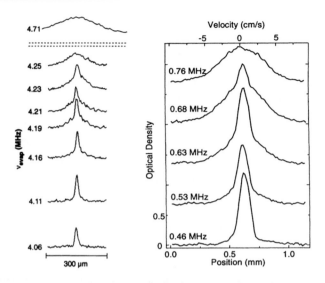

Fig. 3.2 Time-of-flight absorption images of an ultracold cloud for different values of the final evaporation radiofrequency. From top to bottom, the images show the onset of atoms at lower velocities than the one due to a thermal distribution, until they constitute the near totality of the imaged atoms. Left plots are for ^{87}Rb at JILA, right plots for ^{23}Na at MIT. Credits: (Left) reprinted figure with permission from [3]. Copyright 1995 by the American Association for the Advancement of Science. (Right) reprinted figure with permission from [13]. Copyright 1995 by the American Physical Society

region around the center. The latter was more accurately fittable with an inverted parabola in line with the expectations from the Thomas-Fermi profile in a harmonic trap. By lowering even further the minimum frequency for evaporative cooling, only the inverted parabola was observed, with no discernible thermal component, as expected for a pure condensate.

Two important comments are in order. First, in principle, a Bose condensate has no associated temperature; its distribution does not contain any dependence on temperature. The term "temperature of the condensate" is an abbreviation to indicate the temperature of the thermal cloud, in the nondegenerate regime, also present in the trap. While in the trap, this temperature is measured by fitting the density profile of the cloud in the regions in which this is manifestly expressed by a Gaussian function, outside the Thomas-Fermi region. As the temperature decreases, the thermal component decreases, and the assessment of its temperature becomes more problematic, being ultimately limited by the weak observed signal in the absorption profile. This issue is exacerbated in the case of fermions, as we will comment later, due to the absence of a sharp phase transition unless the fermions undergo bosonization. Second, measurement of the condensate fraction allows to check the dependence upon temperature and to compare with the prediction, in a harmonic trap, as in Eq. (1.111). However, it should be kept in mind that even in principle we cannot claim the interpretation of the data in terms of a phase transition. Apart

Fig. 3.3 Absorption images of an ultracold cloud for different times of flight. The drastic change in the aspect ratio of the cloud is evident, as well as the dimmer isotropic shape of the thermal component at later times. Credit: reprinted figure with permission from [30]. Copyright 1996 by the American Physical Society

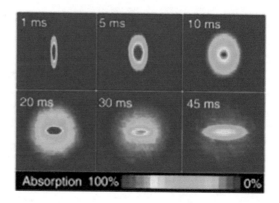

from the experimental errors in determining temperature and condensate fraction, with a given (necessarily finite and in general small, order of 10^9 at most) number of atoms, there is a smooth dependence of the condensate fraction on temperature, without the formation of the kink expected at the transition temperature for ordinary phase transitions involving Avogadro numbers of atoms or molecules [23]. This demystifies a bit Bose-Einstein condensation as a dramatic phenomenon, in practice it has more the nature of a crossover between two regimes, and in doing this, it shares the analogous behavior of fermions in the noninteracting approximation.

Analogous evidences were reported from the MIT group led by Wolfgang Ketterle using ^{23}Na [13]. In this case, the Majorana losses were avoided using a blue-detuned laser beam focused on the center of the magnetic trap, a sort of optical plug, with the usual MOT precooling stage, preceded by a Zeeman slower. The use of the Zeeman slower and the large lifetime of the atoms in the trap allowed for large atom numbers. Later on, the trap was replaced by a Ioffe-Pritchard trap yielding even larger numbers of atoms with the further advantage of creating condensates with variable aspect ratio. In Fig. 3.2, we show the original plots of the absorption images after a fixed time of flight for different values of the final radiofrequency for forced evaporative cooling for the JILA (left) and MIT (right) experiments.

The flexibility offered by the Ioffe-Pritchard traps was key to evidence the dynamics of the aspect ratio for the Bose condensed and the thermal components. If a nondegenerate atomic cloud is trapped in a highly anisotropic potential, with large aspect ratio for the corresponding spatial shape of the cloud, we expect a nearly isotropic cloud after a time of flight much larger than the reciprocal of the trapping frequencies, since the momentum distribution in the cloud is uncorrelated with the spatial profile at thermal equilibrium. In the opposite case of a Bose-condensed cloud, we instead expect the inversion of the aspect ratio because the higher spatial confinement implies a higher value of momentum in the same direction, as a consequence of the uncertainty principle. Therefore, after some time of flight, a Bose-condensed cloud becomes more elongated in the direction in which the confinement in the preexisting trap was stronger. This dramatic change in aspect ratio of the Bose condensed component is what is manifest in Fig. 3.3 in the case

3.2 First-Generation Experiments

of the MIT experiment, while it may be noticeable that the lower density thermal component in the periphery is isotropic [30].

More quantitative evidences have been collected by measuring the internal energy of the cloud, related to the speed of expansion after switching off the trap, and its scaling with the number of atoms. This allows comparison with the predictions of the Gross-Pitaevskii equation, in principle yielding a measurement of the elastic scattering length. With the use of Feshbach resonances [20], a further knob has become available via modulation of the elastic scattering length, with the possibility to explore fast expansions of the cloud simulating some aspects of supernovae events, the so-called bosenovae. However, the use of Feshbach resonances in this context has been limited, in the case of Bose gases, by the concomitant three-body losses and following heating of the sample [40]. As we will describe later, these losses are far less important for fermions, and indeed, Feshbach resonances have been a crucial tool to study superfluidity.

Another important test of mean-field theory consists in the study of the propagation of density perturbations of the condensate [5]. This has been achieved by using dispersive imaging, see Fig. 3.4, and a blue-detuned laser beam focused on the center of the trap suddenly switched on or off to create a perturbation in the Thomas-Fermi profile of the condensate. By taking various dispersive images of the cloud while still in the trap, the speed of sound can be measured and related to the density in the center of the cloud prior to the application of the laser beam, providing a test of the square root dependence of the sound speed $c_s(r) = (4\pi \hbar^2 a n(r))^{1/2}/m$

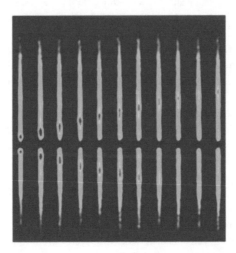

Fig. 3.4 Observation of sound waves in a Bose-Einstein condensate. Eleven consecutive phase-contrast images of the condensate profile, spaced by 1.3 ms intervals, are shown after a density perturbation in the center was suddenly created with a blue-detuned laser beam, 1 ms before the first image. The perturbation moves at constant speed along the condensate profile, with progressive spreading. Credit: reprinted figure with permission from [5]). Copyright 1997 by the American Physical Society

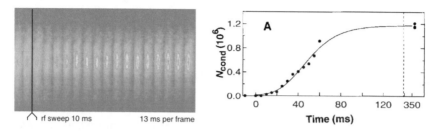

Fig. 3.5 Nondestructive images of the condensate formation after a fast evaporative quench and evidence for an initial exponential growth of the condensate. Credit: reprinted figures with permission from [31]. Copyright 1998 by the American Physical Society

without free parameters, once the density is also measured. In addition, the density perturbation was also observed to spread during the propagation, ruling out solitonic excitations.

The confinement of Bose-Einstein condensation in optical dipole traps opened also the possibility to study multicomponent Bose condensates, the so-called "spinor" condensates, and their properties of miscibility or phase separation [18,39]. This research direction has evolved in time to constitute a large bulk of experimental and theoretical studies of the Bose condensates with various internal symmetries, aimed at testing many-body quantum models with a controllability unparalleled in condensed matter physics, with magnetic and superfluid properties simultaneously present. In these exploratory stage, mean-field theory was verified, but then soon, as we will discuss in Chap. 4, the most ambitious task of exploring regimes in which mean-field theory fails was tackled, most importantly in the study of quantum phase transitions, atoms in lower dimensionality, and dynamical many-body phenomena.

First studies also allowed to evidence crucial coherence properties of the Bose condensates which are rather independent upon the many-body interactions. The dynamics of formation of a condensate was studied after a sudden evaporation using phase-contrast imaging [31]. This allowed to evidence the presence of an exponential grow of atoms in the condensate at early times, interpreted as a signature of lasing amplification, as seen in the right plot in Fig. 3.5.

However, the more convincing evidence for the intrinsically coherent nature of the condensate was achieved by studying their interference [6]. After creation of an elongated condensate of sodium atom, the MIT group succeeded in splitting the condensate into two halves by using a blue-detuned laser beam focused on its center and released this configuration of two blobs of condensate from the magnetic trap. The condensates in free fall were then imaged in a horizontal plane and interference fringes with the expected spacing, and pattern was observed with absorption imaging. Various consistency tests, including dropping a single condensate after blasting the other, were also performed. These studies led to a growing interest for using Bose condensates in metrological applications, for instance, as input states of atomic interferometers. As we will describe in Chap. 5, this research direction also includes the possibility of circuits employing atomic

3.2 First-Generation Experiments

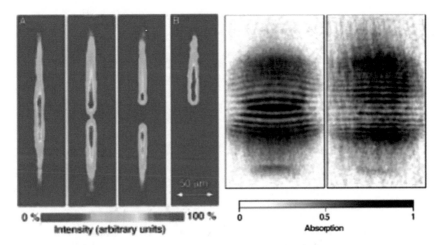

Fig. 3.6 Observation of coherence of a Bose-Einstein condensate. (Left) four in situ absorption images of the Bose condensate with various optical manipulations, from left to right the whole condensate, two separated condensates with different average spacing and a single half of the condensate after blasting the other half with a laser pulse on resonance. (Right) time-of-flight absorption images in the horizontal plane after a split condensate is released from the trap, with two different initial separation of the two clouds, resulting in different spatial periodicity of the interference fringes. Credit: reprinted figure with permission from [6]. Copyright 1997 by the American Association for the Advancement of Science

superfluid junctions (atomtronics) as well as the use of ultracold atoms as platforms for quantum computation.

Various forms of output couplers for Bose condensates, also named "atom lasers," have been developed since the pioneering MIT study in which the output coupler consisted of a sequence of atomic pulses with a constant repetition time, the analog of a pulsed laser. In particular, coherent interference of atoms falling under gravity after release from an optical lattice, using Raman transitions, and continuous-wave couplers have been demonstrated (Fig. 3.6).

The first evidences for quantum degeneracy of fermions were reported by the Deborah Jin group at JILA in 1999 [14]. Fermions are more difficult to deal with due to a variety of reasons, both accidental and fundamental. Due to nuclear pairing, there are only two stable fermionic isotopes among alkali atoms, ^{6}Li and ^{40}K, and their hyperfine structure makes sub-Doppler cooling less efficient. There is no sharp phase transition while cooling down fermions; their behavior is of a crossover nature, with a smooth morphing from the classical case to the quantum degenerate case. Finally, fermions must obey the Pauli "principle," therefore collecting around a single point identical fermions at high density and in the same spin state becomes progressively challenging, a phenomenon also known like "Pauli blocking" (Fig. 3.7).

Based on the considerations above, the initial evidences for fermion degeneracy have been far more circumstantial than in the case of bosons. The JILA group

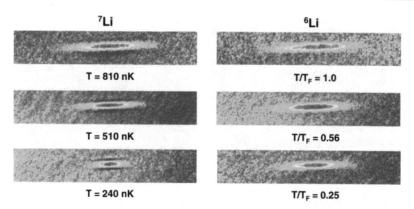

Fig. 3.7 Evidence for Pauli pressure in fermionic lithium. On the left, absorption images of bosonic ^{7}Li are shown at different temperatures decreasing from top to bottom. The temperature on the top images corresponds to the Fermi temperature of the ^{6}Li cloud. Notice that the size of the bosonic and fermionic clouds on the top images are basically indistinguishable. As the temperature is decreased, the bosonic clouds shrink more than its fermionic counterpart, as visible especially when the fermions reach a degeneracy parameter $T/T_F = 0.25$. A quantitative analysis of the images confirms that the size of the Fermi cloud reaches a constant value, as expected by a temperature-independent outer pressure consequence of the Pauli principle. Credit: reprinted figure with permission from [42]. Copyright 1997 by the American Association for the Advancement of Science

evidenced a decreased efficiency in dual evaporative cooling of two hyperfine states of ^{40}K, which is expected based on Pauli blocking, but one cannot rule out other sources of inefficiency in cooling atoms peculiar to ultracold temperatures. The direct imaging of the clouds allowed the group at Rice University led by Randall Hulet to evidence broader imaging profiles for fermionic ^{6}Li having as a comparison imaging profiles for bosonic ^{7}Li, under similar trapping conditions [42]. This has been interpreted as due to the Pauli pressure among fermions, the opposite phenomenon to the pileup expected for quantum degenerate clouds of bosons.

The possibility to tune interactions via Feshbach resonances has resulted more crucial for fermions, revealing favorable surprises. Superfluidity of fermions, as already known from the electron case in superconductors, relies on an effective attraction in the many-body limit. In the case of atoms, there is no need to overcome the Coulomb repulsion present in the electron case, and effective attractive interactions are generated if the elastic scattering length is negative. A crucial finding was that, unlike the case of bosons, Feshbach resonances in Fermi gases are not accompanied by large inelastic interactions, so atom losses are minimal. This allows, as discussed in detail in the next chapter, to study quantum transport phenomena in the entire range of effective interatomic interactions, morphing from a regime of superfluidity of tightly bound pairs of fermions to the regime of loosely bound fermions in an intrinsically many-body state, with the average distance between the two fermions much larger than the average distance between any two fermions. By tuning the magnetic field close to a Feshbach resonance, one can also study a regime of extremely strong interactions, in which universal properties emerge regardless of the microscopic details of the Fermi atoms used, allowing for analog studies of similar systems already identified, for instance, in astrophysics, like quantum degenerate stars.

Initial studies of strongly interacting Fermi gases focused on the hydrodynamic behavior of a degenerate Fermi gas after its release from the trap or studying collective excitations in situ [25, 34]. Precision thermometry allowed to measure the heat capacity of a strongly interacting Fermi gas [25]. Precision thermometry of fermions in the quantum degenerate regime is difficult, unless they coexist at thermal equilibrium with a Bose gas with a substantial thermal component. In the quantum degenerate regime, the average number of fermions per state is quite insensitive to temperature, as it is well approximated by a Heaviside function. This makes the study of their phase diagram, which will be crucial in studying model Hamiltonians emulating condensed matter systems manifesting temperature-driven phase transitions, more difficult. In addition, strongly interacting Fermi gases with repulsion interatomic interactions were produced showing that they form effective bosonic dimers which are available to Bose condense [38, 44]. Radiofrequency spectroscopy also provided evidences for the existence of a pairing gap, an important feature of Cooper pairing [11].

Ultimately, firm signatures of the achievement of the degenerate regime became available by studying transport properties and the onset of superfluidity. This was obtained with a technological tour de force, both in increasing the number of atoms available in the experiments and using Feshbach resonances to tune the interatomic interactions.

Studying transport properties requires considerable number of atoms, for instance, to imprint vortices and vortex lattices, and the dual evaporative cooling of two hyperfine states of fermions is not optimal in this regard due to the progressively decrease number of atoms. This is affecting the possibility to achieve robust signals as well as in decreasing the Fermi energy of the sample. For this reason, several experiments focused on sympathetic cooling between a Bose species present in large amount and the Fermi species. This implies studies of interspecies elastic scattering, as thermalization between the two samples becomes difficult if the interaction between bosons and fermions is too weak. Several mixtures have been used, in particular the MIT group succeeded in bringing a large number of fermions to quantum degeneracy by using a ^{6}Li-^{23}Na mixture [16].

It should also be mentioned that precision thermometry of fermions in the quantum degenerate regime is difficult, unless they coexist at thermal equilibrium with a Bose gas with a substantial thermal component. In the quantum degenerate regime, the average number of fermions per state is quite insensitive to temperature, as it is well approximated by a Heaviside function. This makes the study of their phase diagram, which will be crucial in studying model Hamiltonians emulating condensed matter systems manifesting temperature-driven phase transitions, more difficult.

3.3 Quantum Transport Properties of Weakly Interacting Bose Gases

A distinctive feature of degenerate quantum gases is their response to external forces or torques and the consequent transport properties. This class of phenomena can be understood, in the case of Bose gases, using the weakly interacting approximation, and allows to capture most of, but not all, the physics of superfluid ^{4}He. At the same time, this can be considered as a pristine example of nonstationary phenomenon, thereby requiring the time-dependent Gross-Pitaevskii equation Eq. (1.129), which can be rewritten in terms of hydrodynamical variables, introducing one scalar field and one vector field, density and velocity of the condensate, respectively:

$$\Psi(\mathbf{r}, t) = n(\mathbf{r}, t)^{1/2} e^{i\theta(\mathbf{r},t)} . \tag{3.4}$$

In this way, Eq. (1.129), after separating the real and imaginary parts, is rewritten as two differential equations in the real domain:

$$\frac{\partial}{\partial t} n + \nabla \cdot (n\,\mathbf{v}) = 0 , \tag{3.5}$$

$$m\frac{\partial}{\partial t}\mathbf{v} + \nabla \left[\frac{1}{2} m v^2 + U(\mathbf{r}) + gn - \frac{\hbar^2}{2m} \frac{\nabla^2 \sqrt{n}}{\sqrt{n}} \right] = \mathbf{0} . \tag{3.6}$$

The first equation has the form of a continuity equation, in this case expressing the conservation of the total number of atoms. This allows to identify the velocity field as

$$\mathbf{v}(\mathbf{r}, t) = \frac{\hbar}{m} \nabla \theta(\mathbf{r}, t). \tag{3.7}$$

Notice that, being expressed as the gradient of the phase, the velocity field is irrotational $\nabla \wedge \mathbf{v} = \mathbf{0}$ and, consequently, by evaluation the circulation of the velocity field $\mathbf{v}(\mathbf{r}, t)$ along any closed contour $\mathcal{C}$, we get

$$\oint_{\mathcal{C}} \mathbf{v} \cdot d\mathbf{r} = \frac{\hbar}{m} \oint_{\mathcal{C}} \nabla \theta \cdot d\mathbf{r} = \frac{\hbar}{m} \oint_{\mathcal{C}} d\theta = k\frac{h}{m}, \tag{3.8}$$

with k an integer number. This implies quantization of the circulation in integer multiples of h/m. As we will discuss soon, this property determines a peculiar response of the Bose condensate when angular momentum is imparted and leads to the generation of quantized vortices, without a classical counterpart.

The second equation is the local, hydrodynamic form of the Newton's equation for a fluid in the presence of the kinetic energy and the external potential energy for each atom, a nonlinear term expressing the relation between the interatomic energy and the density, i.e., the equation of state of the Bose gas, and a nonlinear quantum

3.3 Quantum Transport Properties of Weakly Interacting Bose Gases

correction term of second order in the Planck constant, depending on the spatial distribution of the atomic density.

In the absence of the interatomic term, for $g = 0$, and for just one particle, Eqs. (3.5)–(3.6) are known as Madelung equations, completely equivalent to a single-particle Schrödinger equation. For this reason, a many-body system described by these equations in general is also named Madelung fluid.

The main advantage of the hydrodynamic picture of the Gross-Pitaevskii equation is that it allows for a close comparison with a viscous fluid having a generic zero-temperature equation of state $\mu(n)$:

$$\frac{\partial}{\partial t} n + \nabla \cdot (n\mathbf{v}) = 0 \tag{3.9}$$

$$m\frac{\partial}{\partial t}\mathbf{v} + \nabla\left[\frac{1}{2}mv^2 + U(\mathbf{r}) + \mu(n)\right] = \eta\nabla^2\mathbf{v} + m\mathbf{v} \wedge (\nabla \wedge \mathbf{v}) \tag{3.10}$$

where Eq. (3.9) is the usual continuity equation and η in Eq. (3.10) is the viscosity coefficient. Any generic fluid in the collisional regime satisfies these equations, and Eq. (3.10) is also known as Navier-Stokes equation.

The comparison shows that the Gross-Pitaevskii hydrodynamics differs from the one of a Navier-Stokes fluid by the presence of a genuine quantum term $-\hbar^2\nabla^2\sqrt{n}/(2m\sqrt{n})$, expressing the propensity of the wave function to spread out to reduce the contribution of the associated kinetic energy. Moreover, the viscosity term is absent in the former equation, and, as remarked earlier, the velocity field $\mathbf{v}(\mathbf{r}, t)$ is irrotational, unlike the case of the classical fluid for which a rigid rotation, with all components moving with uniform angular velocity, is possible.

The absence of the viscosity term may suggest that the Gross-Pitaevskii equation contains dissipationless motion of the fluid, reminiscent of the superfluidity of ^{4}He. Likewise, an irrotational velocity field may indicate that the condensate will not respond to the application of an external torque, in such a way that no angular momentum can be imparted. This requires more in-depth scrutiny, as we will find out that there are thresholds for both the linear velocity and the angular velocity above which the response of the condensate no longer follow these expectations.

In order to clarify the situation, we discuss how a Bose condensate can trade energy, linear momentum, and angular momentum with the external environment. Let us consider Eqs. (3.5) and (3.6) assuming for simplicity $U(\mathbf{r}) = 0$. We now discuss the case of small perturbations in the density and velocity fields for a condensate initially at rest:

$$n(\mathbf{r}, t) = n_{eq} + \delta n(\mathbf{r}, t), \quad \mathbf{v}(\mathbf{r}, t) = \mathbf{0} + \delta\mathbf{v}(\mathbf{r}, t), \tag{3.11}$$

where $\delta n(\mathbf{r}, t)$ and $\delta\mathbf{v}(\mathbf{r}, t)$ represent small variations with respect to the stationary configuration with density n_{eq}.

By neglecting quadratic terms in the variations, from Eqs. (3.5) and (3.6), we get

$$\frac{\partial}{\partial t}\delta n + n_{eq}\nabla \cdot \delta \mathbf{v} = 0 , \qquad (3.12)$$

$$\frac{\partial}{\partial t}\delta \mathbf{v} + \frac{c^2}{n_{eq}}\nabla \delta n - \frac{\hbar^2}{4m^2 n_{eq}}\nabla(\nabla^2 \delta n) = \mathbf{0} . \qquad (3.13)$$

By applying the time derivative operator to Eq. (3.12) and the divergence to Eq. (3.13), we obtain

$$\left[\frac{\partial^2}{\partial t^2} - c_s^2 \nabla^2 + \frac{\hbar^2}{4m^2}\nabla^4\right]\delta n(\mathbf{r}, t) = 0 , \qquad (3.14)$$

where we have introduced a quantity with the dimensions of velocity, c_s, satisfying the relationship

$$c_s^2 = \frac{g\, n_{eq}}{m} . \qquad (3.15)$$

The first two terms in Eq. (3.14) represent wave propagation, with velocity c_s, of a density perturbation. This is the analog, for a Bose condensate, of the sound velocity in a classical fluid at nonzero temperature and is therefore called zero-temperature sound velocity of the Bose gas. Equation (3.14) admits monochromatic plane-wave solutions but, unlike usual plane waves, the presence of the interaction term, the third term in the equation, does not allow for the usual dispersion relation between the frequency ω and the wavevector $\mathbf{q}$ for a classical gas, $\omega(q) = c_s q$, with c_s the sound speed, corresponding to the propagation of phonons when second quantization is considered. Plane waves are however allowed for a Bose gas provided that a dispersion relationship $\omega = \omega(q)$ is satisfied as

$$E(q) = \hbar\,\omega(q) = \sqrt{\frac{\hbar^2 q^2}{2m}\left(\frac{\hbar^2 q^2}{2m} + 2mc_s^2\right)} . \qquad (3.16)$$

This is the so-called Bogoliubov dispersion relation for the elementary excitations in a weakly interacting Bose gas at zero temperature [7]. In analogy to usual phonons as quantized, collective excitations of density perturbations in fluids or solids, these excitations are also named Bogoliubov phonons.

Notice, see Fig. 3.8, that the Bogoliubov dispersion is well approximated by the phonon dispersion relation $\omega(q) = c_s q$ in the regime $q \ll 2mc_c/\hbar$, for density perturbations with large wavelength. In the opposite regime, the dispersion relation becomes the one of a free particle. Equation (3.16) is crucial because any amount of energy and momentum effectively imparted to the Bose gas must always follow the dispersion relation curve. In this way, we can think that a weakly interacting Bose gas can be replaced, in regard to the response to weak perturbations from the stationary state, by a noninteracting Bose gas plus a gas of elementary excitations,

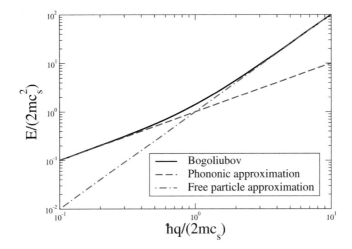

Fig. 3.8 Dispersion relation for elementary excitations in a weakly interacting Bose gas. The energy of the elementary excitation, in units of $2mc_s^2$, is plotted versus its momentum in units of $2mc_s$. With this choice, the dimensionless dispersion relation becomes $y = x^2\sqrt{1 + 1/x^2}$, having defined $E/(2mc_s^2) = y$, $\hbar q/(2mc_s) = x$. Also shown are the limits for $x \to 0$ and $x \to \infty$, the dispersion relationships for Bogoliubov phonons and for free particles, respectively, intersecting each other at $x = 1$

the Bogoliubov phonons, trading energy and momentum between the Bose gas and the external environment.

3.3.1 Translational Response of a Bose-Einstein Condensate

Let us consider a macroscopic particle of mass M that is moving in the superfluid with velocity $\mathbf{v}$. We want to discuss the possibility of energy and momentum exchange between the macroscopic particle and the superfluid, assuming that this can only happen through creation of quasiparticles. Then, the conservation of linear momentum and energy for the macroscopic particle of mass M and one quasiparticle of the superfluid is written as

$$\frac{1}{2}Mv^2 = \frac{1}{2}M(v')^2 + E(p), \quad M\mathbf{v} = M\mathbf{v}' + \mathbf{p}, \tag{3.17}$$

where $\mathbf{v}$ and $\mathbf{v}'$ are the velocities of the macroscopic particle before and after the quasiparticle creation and $E(p)$, $\mathbf{p}$ are energy and momentum of the quasiparticle, related via Eq. (3.16). Combining the two equations above, one finds

$$\mathbf{v} \cdot \mathbf{p} = vp\cos\alpha = E(p) + \frac{p^2}{2M}. \tag{3.18}$$

where $\cos\alpha = \hat{\mathbf{v}} \cdot \hat{\mathbf{p}}$. In the limit of a macroscopic mass $M \to \infty$, we can omit the second term on the right-hand side and therefore

$$v = \frac{E(p)}{p\cos\alpha}. \tag{3.19}$$

Since $|\cos(\alpha)| \leq 1$, there is a lower bound for the modulus of the velocity v, and a necessary condition for the existence of energy and momentum exchange is that v is larger than the minimum of $E(p)/p$ for any choice of p. This identifies a critical velocity for dissipationless motion, also called Landau critical velocity:

$$v_c = \min_{\mathbf{p}}\left(\frac{E(p)}{p}\right). \tag{3.20}$$

Only for velocities $\mathbf{v}$ larger in modulus that v_c it is possible to have energy exchange between the macroscopic object and the Bose condensate. In the case of the Bogoliubov dispersion relation, this minimum is $v_c = c_s$. In the case of a free-particle dispersion relationship, obtained for a noninteracting Bose condensate, one gets $v_c = 0$. Superfluidity therefore is not present for ideal, noninteracting, Bose-Einstein condensates. Also, the fact that this is a necessary condition for superfluidity does not imply that it is sufficient. In general, there can be other ways to exchange energy other than the creation of microscopic excitations in the form of quasiparticles.

3.3.2 Rotational Response of a Bose-Einstein Condensate

We have discussed the quantization of circulation of the velocity field as in Eq. (3.8). If the integer number $k = 0$, no rotation can be imparted to the condensate. If instead $k \neq 0$, the condensate as a whole will still be at rest, but there will be points inside a closed path $\mathcal{C}$ supporting the presence of angular momentum. This implies that the domain where $\mathbf{v}$ is well defined is multiply connected, and there will be j points, with $j \leq k$ (called defects) in the circulation field which cannot be avoided by an appropriate change in the closed path, i.e., they are topological defects. For a multiply connected domain $\mathcal{D}$, with $\mathbf{r} \in \mathcal{D}$, one gets $\nabla \wedge \mathbf{v}(\mathbf{r}) = \mathbf{0} \implies \mathbf{v}(\mathbf{r}) = \nabla \chi(\mathbf{r})$ with $\chi(\mathbf{r})$ a multivalued scalar field. A simple example of topological defect is a quantized vortex along the z axis, with number density in the $x - y$ plane:

$$n(r) \simeq n_\infty \frac{r^2/\xi^2}{1 + r^2/\xi^2} \quad \text{and} \quad v(r) = k\frac{\hbar}{m}\frac{1}{r}, \tag{3.21}$$

where $n(r)$ is the density in the radial plane, only depending on the cylindrical radial distance $r = \sqrt{x^2 + y^2}$ from the origin, n_∞ is the density at infinite distance, ξ is the healing length defined in Eq. (1.133), and $v_s(r)$ is the magnitude of the velocity field at distance r from the origin. Notice the singularity of the velocity field at $r = 0$

3.3 Quantum Transport Properties of Weakly Interacting Bose Gases

and the fact that the density drops by a factor 2 if $r = \xi$. The healing length, already introduced when discussing the Thomas-Fermi approximation in Chap. 1, therefore measures the "size" of the vortex, over which the density significantly changes. The integer number k is called charge (or topological charge) of the vortex, which can be negative or positive, depending on the direction of the corresponding angular momentum.

We can evaluate the total energy associated to a vortex in the approximation of a constant number density in between a minimum distance r_0 and a maximum distance R_0 from the origin of the vortex and with zero density elsewhere, corresponding to the macroscopic wave function:

$$\psi(r, \phi) = \psi_0 \, e^{ik\phi} \quad (r_0 \leq r < R_0, \; 0 \leq \phi < 2\pi). \tag{3.22}$$

Indicating the number density as $n = |\psi_0|^2$ and the velocity at distance R as $v = (\hbar/m)(k/r)$, we obtain the vortex energy, apart from a space-independent constant, as

$$E_k = \int n \frac{1}{2} m \, v^2 \, d^2\mathbf{r} = k^2 \frac{\hbar^2}{2m} \int_{r_0}^{R_0} dr \, r \int_0^{2\pi} d\phi \frac{1}{r^2} = k^2 \frac{\pi \hbar^2 n}{m} \ln\left(\frac{R_0}{r_0}\right). \tag{3.23}$$

The minimum distance r_0 can be taken to be the healing length ξ. Since ξ scales as $g^{-1/2}$, we expect vortices to disappear in the case of a noninteracting Bose-Einstein condensate. Therefore, an ideal Bose-Einstein condensate does not have frictionless motion and does not support quantized vortices, confirming the physical intuition of Landau about the fact that superfluidity is not simply reducible to Bose-Einstein condensation (Fig. 3.9).

3.3.3 Experiments on Critical Velocities and Quantized Vortices

Frictionless motion has been evidenced by stirring back and forth at constant velocity a blue-detuned beam into the Bose condensate and comparing the thermal component at the beginning and the end of the stirring stage. It has been evidenced that the condensate fraction depends on the velocity of the stirring beam with an initial plateau and then taking off linearly at higher velocities [36]. The critical velocity was estimated to be about half of the peak sound speed in the condensate. Comparative calorimetric measurements with a thermal cloud confirmed the absence of a heating threshold for this case, and further evidences were obtained by studying the effect of the drag force expected about the critical velocity on the profile of the condensate using dispersive imaging [35]. These experimental studies were confronted with already existing numerical simulations on the behavior of a Bose condensate in the presence of a macroscopic obstacle [15,19,21,43], in which vortex shedding was found to be responsible for the dissipation below the Landau critical velocity. Microscopic obstacles, with size much smaller than the healing length of

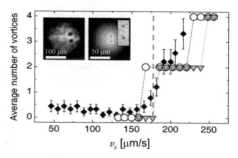

Fig. 3.9 Vortex shedding in a Bose-Einstein condensate moving around a macroscopic obstacle. (Left) numerical simulations showing the two-dimensional profile of the modulus of the macroscopic wave function for a Bose-Einstein condensate in the presence of a cylindrically shaped object with sharp boundaries. Two vortices are symmetrically generated in proximity of the obstacle, on its right side. (Right) observation of vortex shedding. The plot presents the observed average number of vortices versus the velocity of the blue-detuned beam (black diamonds, averages over ten measurements), as well as the expectations from numerical simulations with various parameters, the actual number of atoms (2×10^6) and temperature (52 nK) (triangles joined with dotted lines), and a smaller number of atoms (1.3×10^6 and zero temperature (open circles). The vertical dashed line indicates the estimated critical velocity, and the insets show a vortex pair observed (left) and simulated (right). Notice, at higher velocities, the formation of multiply charged vortex pairs and the stepwise shape of the numerical simulations. Credit: (Left) reprinted figure with permission from [15]. Copyright 1992 by the American Physical Society. (Right) reprinted figure with permission from [32]. Copyright 2010 by the American Physical Society

the condensate, were created in the form of "impurity" atoms inside the condensate, and the corresponding critical velocity for scattering was instead found to be equal to the sound speed in line with the expectation from the Landau criterion [10]. The ultimate proof of vortex shedding was achieved in the observation, always by stirring a blue-detuned beam, of quantized vortex dipoles in highly oblate condensates, i.e., vortex pairs with opposite charges and high stability against annihilation [32].

The study of critical velocities has required several years of refinements, due to the many concomitant sources of heating and the intrinsic inhomogeneous nature of the Bose-Einstein condensates in harmonic traps, with a more recent example of quality data available in [26]. Instead, far more immediate evidences of the response of a condensate to rotational motion have been collected. This is even more relevant because, vice versa, in liquid ^{4}He experiments on critical velocities are relatively simple but the formation and observation of quantized vortices has not been trivial.

Quantized vortices have been successfully generated in Bose-Einstein condensates and their formation and decay studied in detail. Two techniques have been used to imprint angular momentum. The first relies on the imprinting of an angle-dependent quantum phase ϕ through a rotating laser beam, as first demonstrated by the JILA group [29] (see Fig. 2.8). The second instead consists in inducing a mechanical stirring of the condensate using rotating focused laser beams, as performed by the group of Jean Dalibard at the ENS Paris [28]. This group also succeeded in generating a few vortices and in evidencing the various step-like

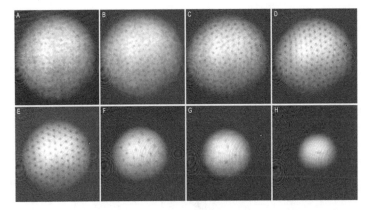

Fig. 3.10 Time evolution of quantized vortices in a Bose-Einstein condensate of ^{23}Na atoms with absorption images after a variable and increasing hold time, showing their formation, stabilization, and progressive decay. The image H on the bottom right is relative to the survival of one vortex after an hold time of 40 s. Credit: reprinted figure with permission from [1]. Copyright 2001 by the American Association for the Advancement of Science

dependence of the number of vortices upon the angular velocity of the stirring beams. Using condensates with larger numbers of atoms, the MIT group generated large number of vortices [1], as shown in Fig. 3.10. Due to the vortex-vortex interaction energy, they organize themselves into regular patterns, the so-called Abrikosov lattices first evidenced in superconductors. The same group has also studied in detail the mechanism of vortex nucleation [37] and the dynamics of formation and decay of vortices at variable temperature [2]. Surface modes were found to play an important role in vortex nucleation, and as expected, the lifetime of vortices strongly depends on temperature.

3.4 Superfluidity in ^{4}He

Superfluidity is indicating a vast class of phenomena, the most surprising one being a zero viscosity flow. The first experimental evidence for a phenomenon strictly related to superfluidity, i.e., superconductivity, later seen as the superfluidity of an electron gas, goes back to 1911. Kamerlingh Onnes in Leiden found a surprising and abrupt drop in the resistivity of solid mercury at a temperature of 4.2 K, obtained by contact with helium which just liquefies, at the atmospheric pressure, at the same temperature. In his own words, "Mercury has passed into a new state, which on account of its extraordinary electrical properties may be called the superconductive state." Thermal and magnetic properties were also investigated, confirming that at the same temperature a discontinuity in the heat capacity as well as a complete expulsion of the magnetic field occur, corroborating the hypothesis that a new state of matter was formed via a phase transition. In the same experiments in Leiden, Kamerlingh Onnes also reached the record temperature of about 1.5 K by reducing

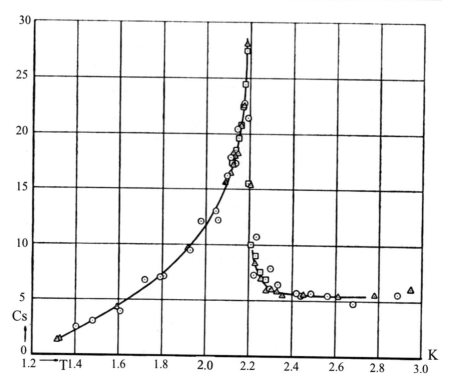

Fig. 3.11 Specific heat of helium 4 as a function of the temperature around the critical temperature T_λ. This critical temperature is called lambda point, from the shape of the curve (from [22])

the pressure of liquid helium. Ironically, it took almost three decades to discover that a phenomenon very similar to superconductivity was already occurring in liquid helium itself. Superfluidity of the most abundant stable isotope of helium, ^{4}He, was discovered in 1937 by Pyotr Leonidovich Kapitsa, John Allen, and Don Misener, who found that below $T_\lambda = 2.16$ K ^{4}He remains liquid but shows zero viscosity. Also, in correspondence of T_λ, the specific heat seems to have a discontinuity and certainly has a maximum, as depicted in Fig. 3.11.

In 1938, Fritz London attempted a first theoretical explanation of the superfluidity of ^{4}He on the basis of Bose-Einstein condensation. Indeed, ^{4}He atoms, with an even number of protons, neutrons, and electron, are effective bosons, and inserting in the formula of the critical temperature T_c of a noninteracting Bose gas density and mass of helium ^{4}He, the resulting T_c is close to 2 K. The "lambda" curve of the specific heat is reminiscent, at least qualitatively, of the dependence of the heat capacity of an ideal Bose gas discussed in Chap. 1; see Fig. 1.4. Furthermore, in spite of the search during the following three decades, the absence of superfluidity for ^{3}He in the same range of temperature as for ^{4}He, even if the two liquids have similar densities, was considered a third strong argument in favor of London's hypothesis, as ^{3}He is an effective fermion, therefore not eligible for Bose-Einstein condensation.

3.4 Superfluidity in ⁴He

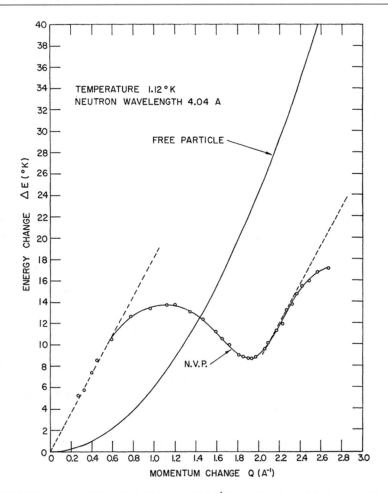

Fig. 3.12 Dispersion relation $E(p)$ of the superfluid ⁴He as a function of the wave number $p/\hbar$, measured via neutron scattering, at the temperature of 1.12 K. The leftmost dashed line indicates the expected dispersion relationship only due to phonon excitations, with the slope being the speed of sound. The Landau critical velocity is predicted to be at the minimum slope of the dispersion curve, corresponding to a wave number of about 1.9 Å^{-1}. Credit: Reprinted figure with permission from [17]. Copyright 1961 by the American Physical Society

However, liquid ⁴He is made of strongly interacting atoms, while Bose-Einstein condensation was understood as a phase transition of noninteracting bosons. Laszlo Tisza first discussed the possibility of the coexistence of two fluids, and Lev Davidovich Landau in 1941 was able to quantitatively describe superfluidity of ⁴He without using Bose-Einstein condensation and, guessing a dispersion relation for elementary excitations, later confirmed experimentally by means of neutron scattering as shown in Fig. 3.12. This dispersion relation is quite different from the Bogoliubov dispersion relation in Eq. (3.16), although at very small momenta contains a linear dependence as expected for phonon excitations. It is worth noting that the Landau critical velocity criterion in this case does not yield a critical velocity equal to the speed of sound c_s, rather a critical velocity v_c smaller by almost one order of magnitude, the slope at the local minimum for the curve interpolating the experimental points in Fig. 3.12. The elementary excitations responsible for the nearly parabolic shape around the local minimum are not present in the case of a dilute Bose gas and are called rotons. Therefore, in liquid helium, there is simultaneous presence of a gas of phonons and rotons, with different dependence upon the temperature. The actual critical velocity measured in liquid ^{4}He is not even c_{min} and depends on the geometry of the container, often resulting in a value lower by one or two orders of magnitude. This has been interpreted by Feynman as due to creation of macroscopic excitations as quantized vortices, showing the the existence of mechanisms alternative to the release of energy via elementary excitations. Some confusion has been created by the terminology, as elementary excitations like rotons, due to their name, seem indicative of a relation to macroscopic excitations like vortices, but this is incorrect. In the dilute Bose gas, there is no rotonic component in the dispersion relationship, yet there are quantized vortices. As we have discussed in the previous subsection, even in the case of Bose condensates, the critical velocity may be subsonic due to the possibility to release energy via vortex shedding.

Another remarkable difference between superfluid ^{4}He and Bose-Einstein condensates of atomic gases is the amount of atoms in the ground state. With neutron scattering, the percentage of atoms in the ground state of superfluid ^{4}He has been measured to be about 8 % at the lowest temperature, while all the atoms are participating to the superfluid state, again confirming that Bose-Einstein condensation and superfluidity are distinct phenomena and that one can exist without the other. Instead, in ultracold bosons, this percentage is close to 100 %, as a result of the much weaker interatomic interactions, and there is no distinction between atoms in the ground state and in the superfluid state, a statement indicating the absence of the so-called *quantum depletion*. The Bogoliubov theory predicts a quantum depletion also in the case of weakly interacting Bose gases, being of order $n_{BEC}/n_0 \simeq 1.5\sqrt{na^3}$. In usual conditions, this quantity is rather small, but increasing the s-wave elastic scattering length near Feshbach resonances, it is possible to evidence the quantum depletion, as performed in [27] for an homogeneous Bose-Einstein condensate.

Within the two-fluid model of Tisza and Landau, below T_λ ^{4}He is characterized by a inviscid superfluid component and a viscous normal component. At zero temperature, only the superfluid component remains, and the equations of superfluid hydrodynamics are

$$\frac{\partial}{\partial t} n_s + \nabla \cdot (n_s \mathbf{v}_s) = 0 , \qquad (3.24)$$

$$m \frac{\partial}{\partial t} \mathbf{v}_s + \nabla \left[\frac{1}{2} m v_s^2 + U(\mathbf{r}) + \mu(n_s) \right] = \mathbf{0} , \qquad (3.25)$$

3.4 Superfluidity in ^{4}He

where $n_s(\mathbf{r}, t)$ is the local superfluid number density, $\mathbf{v}_s(\mathbf{r}, t)$ is the local superfluid velocity, $U(\mathbf{r})$ is the external potential, and $\mu(n_s)$ is the chemical potential of the bulk system. These equations describe extremely well superfluid ^{4}He, ultracold gases of alkali-metal atoms, and also transport phenomena of superconductors.

Equations (3.24) and (3.25) are very similar to Eqs. (3.5) and (3.6): by neglecting the quantum term $-\hbar^2 \nabla^2 \sqrt{n}/(2m\sqrt{n})$ in Eq. (3.6), they become equivalent setting $n = n_s$ and $\mathbf{v} = \mathbf{v}_s$. However, the derivation of Eqs. (3.5) and (3.6) is based on Bose-Einstein condensation and the Gross-Pitaevskii equation, while Eqs. (3.24) and (3.25) are the macroscopic hydrodynamic equations of a viscousless and irrotational fluid.

Moreover, these hydrodynamic equations allow for a generalization at nonzero temperature, in which the viscous normal component plays a role. According to the two-fluid model, at thermal equilibrium, the total number density is $n = n_s + n_n$, where n_s is the superfluid number density and n_n is the normal number density.

At temperature T, in the rest frame of the superfluid, the normal current density of mass is given by

$$\mathbf{j}_n = m n_n \mathbf{v} = \int \frac{d^3 \mathbf{p}}{(2\pi \hbar)^3} \mathbf{p} \, f_B(E(p) - \mathbf{p} \cdot \mathbf{v}), \tag{3.26}$$

where $\mathbf{v}$ is the velocity of the normal fluid and $f_B(E) = [\exp\{E/(k_B T)\} - 1]^{-1}$ is the Bose distribution.

Assuming a small velocity $\mathbf{v}$ and Taylor-expanding the previous formula with respect to $\mathbf{v}$ to the first order, one finds

$$n_n(T) = -\frac{1}{3} \int \frac{d^3 \mathbf{p}}{(2\pi\hbar)^3} \frac{p^2}{m} \frac{d}{dE} f_B(E) = \frac{1}{3 k_B T} \int \frac{d^3 \mathbf{p}}{(2\pi\hbar)^3} \frac{p^2}{m} \frac{e^{\frac{E(p)}{k_B T}}}{\left(e^{\frac{E(p)}{k_B T}} - 1\right)^2}, \tag{3.27}$$

taking into account that $\int d^D\mathbf{p} \, \mathbf{p} \, (\mathbf{p} \cdot \mathbf{v}) F(p) = (1/D) \left(\int d^D\mathbf{p} \, p^2 \, F(p)\right) \mathbf{v}$. Thus, the thermal activation of quasiparticles increases the normal component of the superfluid. The critical temperature T_s of the superfluid-normal phase transition is obtained setting $n_s(T_s) = 0$.

These considerations depend upon spatial dimensionality, as shown by the integration over momenta in Eq. (3.27). In three spatial dimensions, one finds $T_s = T_c$; thus, the critical temperature T_s for the onset of the superfluid-to-normal phase transition coincides the critical temperature T_c for Bose-Einstein condensation. However, in two spatial dimensions, for interacting bosons, $T_s \neq T_c = 0$. Moreover, in one spatial dimension, for interacting bosons, $T_s \neq 0$, while Bose-Einstein condensation does not exist in the first place.

Once again, this confirms that Bose-Einstein condensation and superfluidity in ^{4}He share some analogies but also that in their relationship they are not trivially related by necessary or sufficient conditions. It is also worth to remark, in the same spirit of a comparative analysis, that the experimental evidences for dissipationless motion in Bose-Einstein condensates are not so manifest as in the case of ^{4}He, first of all due to the inhomogeneous geometry of the samples, and more in general for their finite lifetime. There is no analog, for ultracold atoms, of experiments in which superfluid flow persists in annular geometries for long times, a situation even more stricking in the case of persistent electrical currents in superconducting circuits. For a sort of complementarity often found in experimental physics, the production of quantized vortices and the study of their dynamics of formation and decay have instead been far simpler for Bose-Einstein condensates.

3.5 Topological Defects in Lower Effective Dimensionality

The physics of quantized vortices is intrinsically multidimensional since quantities like circulation and angular momentum require to consider at least two-dimensional situations. It is natural to ask if solutions to the Gross-Pitaevskii equation with topological content also exist in one dimension [21]. So far, we have limited the focus on isotropic trapping, but, as discussed in Chap. 2, a variety of traps offer the possibility of anisotropic trapping, at least in the simplest situation of two trapping frequencies equal but quite distinct from the third one. This is particularly feasible with Ioffe-Pritchard magnetic traps, as well as single-beam optical dipole trap. In this hybrid situation, the trap preserves cylindrical symmetry, allowing for simplifications yet giving rise to an effective reduction of the physical dimensionality.

Let us consider the case of harmonic confinement along x and y with angular frequency $\omega_x = \omega_y = \omega_\perp$ and a generic confinement $\mathcal{U}(z)$ along z, namely,

$$U(\mathbf{r}) = \frac{1}{2} m \omega_\perp^2 (x^2 + y^2) + \mathcal{U}(z) . \tag{3.28}$$

This potential admits the separation of variable for the wave function:

$$\psi(\mathbf{r}, t) = \frac{f(z, t)}{\pi^{1/2} a_\perp} \exp\left[-\frac{x^2 + y^2}{2 a_\perp^2}\right] , \tag{3.29}$$

where $f(z, t)$ is the axial wave function and $a_\perp = \sqrt{\hbar/(m\omega_\perp)}$ is the characteristic length of the transverse harmonic confinement. This is an exact result only if the potential $U(\mathbf{r})$ alone is present. By considering also the presence of the mean-field term as in the Gross-Pitaevskii equation, Eq. (3.29) is an Ansatz, approximately valid as far as the harmonic energy term in the x–y plane prevails over the mean-field energy term, therefore in the limit of large $\omega_\perp$. Alternatively, we can see this Ansatz as valid whenever the healing length of the Bose condensate is much larger than the characteristic length of the transverse harmonic confinement, $\xi \gg a_\perp$. In this way, the mean-field term along the $x - y$ plane only acts as a spatially independent term. By inserting Eq. (3.29) into the Gross-Pitaevskii action and

3.5 Topological Defects in Lower Effective Dimensionality

integrating along x and y, the resulting effective action functional depends only on the wave function $f(z, t)$, satisfying

$$i\hbar \frac{\partial}{\partial t} f(z,t) = \left[-\frac{\hbar^2}{2m} \frac{\partial^2}{\partial z^2} + \mathcal{U}(z) + \gamma |f(z,t)|^2 \right] f(z,t), \quad (3.30)$$

where $\gamma = gN/(2\pi a_\perp^2)$ is the effective one-dimensional (1D) interaction strength, and an additive constant depending on $\hbar\omega_\perp$ has been omitted because it does not affect the dynamics. Notice that γ, even dimensionally, differs from the original interaction strength of the full three-dimensional system. The validity of the Ansatz for large $\omega_\perp$, therefore small $a_\perp$, makes the mean-field effects prominent in the z direction. Equation (3.30) is the time-dependent 1D Gross-Pitaevskii equation.

In the absence of axial confinement, i.e., $\mathcal{U}(z) = 0$, the 1D Gross-Pitaevskii equation admits solitonic solutions in the generic form:

$$f(z,t) = \phi(z - vt) \exp\left[\frac{i}{\hbar} \left(mvz - \frac{mv^2}{2}t - \mu t \right) \right], \quad (3.31)$$

with a time-invariant axial density profile ϕ, which can be assumed a real function by reabsorbing any constant phase factor in the exponential term, v is the velocity of propagation, and μ the chemical potential. The Gross-Pitaevskii equation can then be cast in the form of a time-independent equation in the variable $z - vt = \zeta$:

$$-\frac{\hbar^2}{2m} \frac{d^2\phi(\zeta)}{d\zeta^2} + \gamma \phi(\zeta)^3 = \mu \phi(\zeta), \quad (3.32)$$

This equation can be rewritten in the form:

$$\frac{d^2\phi}{d\zeta^2} = -\frac{\partial W(\phi)}{\partial \phi}, \quad (3.33)$$

where

$$W(\phi) = \frac{m\mu}{\hbar^2} \phi^2 + \frac{1}{2} \frac{m\gamma}{\hbar^2} \phi^4. \quad (3.34)$$

Thus, $\phi(\zeta)$ can be seen as the "coordinate" for a fictitious particle at "time" ζ, moving in conservative potential $W(\phi)$. This allows to solve the equation by considering the integral of the motion, the "total energy" in this mechanical analogy

$$K = \frac{1}{2}\left(\frac{d\phi}{d\zeta}\right)^2 + W(\phi) \text{ and, inverting the relation, } \frac{d\phi}{\sqrt{2(K - W(\phi))}} = d\zeta. \quad (3.35)$$

The integration of Eq. (3.35) proceeds in different ways depending on the sign of γ.

If $\gamma < 0$, we expect a localized state due to the effective attraction between the atoms; therefore, in this case, $\phi(\zeta) \to 0$ as $\zeta \to \pm\infty$, and then, $K = 0$. The solution then gives

$$\phi(\zeta) = \sqrt{\frac{m|\gamma|}{8\hbar^2}} \operatorname{sech}\left[\frac{m|\gamma|}{4\hbar^2}\zeta\right] \quad (3.36)$$

with $\operatorname{sech}[x] = 2/[e^x + e^{-x}]$ and $\mu = -m\gamma^2/(16\hbar^2)$. These solutions are named "bright solitons" as the atoms are concentrated in a specific region of space at any given time, with time-independent spreading of their density. Bright solitons are not topological solutions because their density never becomes equal to zero. However, it should be possible to create bright-vortex solitons that are bright solitons containing a quantized vortex. These bright-vortex configurations have been theoretically predicted but not yet observed in experiments with atoms.

In the case $\gamma > 0$, we expect repulsion between the atoms. The case with velocity $v = 0$ is characterized by a condensate at rest. In this case, we expect a depletion of atoms in a given location, and by assuming that $\phi(z) \to \pm\bar{\phi}$ as $z \to \pm\infty$, with $\mu = \gamma\bar{\phi}^2$

$$K = W(\bar{\phi}) = -\frac{1}{2}\frac{m\gamma}{\hbar^2}\bar{\phi}^4 + \frac{m\mu}{\hbar^2}\bar{\phi}^2 = \frac{m\gamma}{2\hbar^2}\bar{\phi}^4 \quad (3.37)$$

and, integrating Eq. (3.35), we get

$$\phi(z) = \bar{\phi}\tanh\left(\sqrt{\frac{m\gamma}{\hbar^2}}\bar{\phi}\, z\right) \quad (3.38)$$

Therefore, the atoms are everywhere apart from a region in which there is depletion, which is complete at $z = 0$. This solution will manifest itself as an absence of atoms in a given location at all times, a "black" soliton. It is important to stress that with $z \in\,]-\infty, +\infty[$ the wave function of the black soliton is not normalizable. In practice, if the potential $\mathcal{U}(z)$ is actually present, the wave function will fall to zero at large distances. Moreover, the black soliton is a topological configuration because it produces a hole in the atomic cloud at $z = 0$, where the phase of the wave function changes from 0 to π. Such a hole can be displaced elsewhere but cannot be eliminated with a continuous transformation of the wave function, thereby its topological character. In this sense, the black soliton is the 1D analog of 2D and 3D quantized vortices, capturing in the simplest way the concept of conserved topological charge. In Fig. 3.13, we plot the probability density for bright and dark solitons for three values of the nonlinear strength γ.

There are also other dark solitonic solutions with a nonzero minimum and travelling with velocity $v \neq 0$. They are called gray solitons and they are solutions

3.5 Topological Defects in Lower Effective Dimensionality

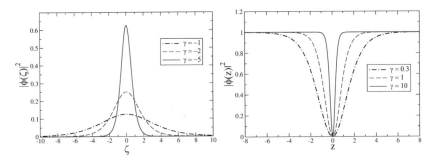

Fig. 3.13 Probability density $|\phi(\zeta)|^2$ of a bright soliton (left plot) and of a black soliton $|\phi(z)|^2$ (right plot), each for three values of the nonlinear strength γ. We set $\hbar = m = 1$

of the time-dependent 1D Gross-Pitaevskii equation. In this case, the wave function $f(z, t)$ is assumed as complex, and it is given by

$$f(z,t) = \sqrt{\bar{f}^2 - f_0^2} \tanh\left[\sqrt{\frac{m\gamma}{\hbar^2}}\sqrt{\bar{f}^2 - f_0^2}(z - vt)\right] + i\, f_0, \quad (3.39)$$

where $\bar{f}$ is the value of the field $f(z, t)$ for $z \to +\infty$ and f_0^2 is the minimum density, related to $\bar{f}$ as $f_0 = (v/c_s)\bar{f}$, with $c_s = \bar{f}\sqrt{\gamma/m}$ the speed of sound. When v approaches c_s, the gray soliton basically morphs into a sound wave, i.e., a travelling wave with a decreasing amplitude and increasing width.

In general, with the words "dark soliton," one indicates a black or a gray soliton, characterized by velocity v such that $0 \leq v \leq c_s$. We also mention the possibility of a vortex bright soliton state, as a hybrid between a vortex (in the x-y plane), and a bright soliton, with wave function:

$$f(x, y, z) = (x^2 + y^2)^{q/2} \exp\left[-(x^2 + y^2)/(2\sigma^2)\right] f(z), \quad (3.40)$$

where q is the topological charge and $f(z)$ the 1D wave function of the bright solution as in Eq. (3.36). Solitons have been found in numerical studies with a choice of $\mathcal{U}(z)$ corresponding to a bistable potential [12]. Stationary solutions were also obtained without a counterpart in the energy eigenstates of the associated Schrödinger problem, and chains of dark or bright solitons were obtained in the limit of strong nonlinearity.

In terms of experimental demonstrations, dark solitons in cigar-shaped Bose-Einstein condensates of ^{87}Rb atoms were created by a phase imprinting method [9]. As visible in Fig. 3.14, an initial imprinted kink then evolved into two kinks moving at a constant speed toward the periphery of the condensate, along the axial direction. The speed of propagation was measured to be significantly smaller than the sound speed, ruling out the propagation as sound waves. In 2002, there were two relevant experiments about bright solitons with BECs made of ^{7}Li atoms,

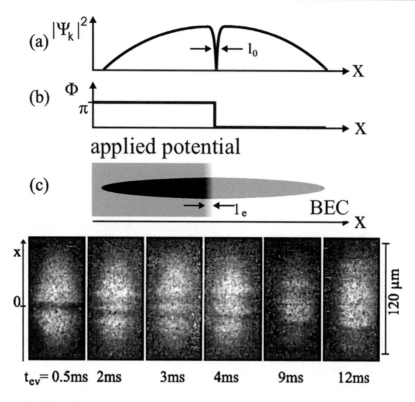

Fig. 3.14 Soliton formation in Bose-Einstein condensates. (Top) density (a) and phase (b) profiles for a dark soliton with a dimple in the density of width l_0 and a π difference in the phase of the macroscopic wave function between the two halves of the condensate, suddenly imprinted with a blue-detuned off-resonant laser beam sent only to the left half of the condensate as in (c). The ^{87}Rb Bose-Einstein condensate is cigar shaped with trapping frequencies of 14 Hz along the axial direction and 425 Hz along the radial direction. (Bottom) absorption density profiles of the condensate at different times of flight t_{ev}, showing the evolution of the initial kink clearly visible after 0.5 ms from the release and splitting into two kinks afterward. Credit: reprinted figures from [9]. Copyright 1999 by the American Physical Society

which has an intrinsic negative scattering length, using Feshbach resonances. Bright solitons were produced in an ultracold ^{7}Li gas by suddenly switching the interatomic interaction from repulsive to attractive before the release of the condensate in a one-dimensional optical waveguide, which is attractive in the transverse direction but repulsive in the longitudinal direction [24]. Propagation of the soliton without dispersion over a macroscopic distance of 1.1 millimeter was observed. With a similar preparation technique, [41] reported the formation of a train of bright solitons of ^{7}Li atoms in a quasi-one-dimensional optical trap. The solitons were set in motion by offsetting the optical potential and were observed to propagate in the longitudinal harmonic potential for many oscillatory cycles without spreading. More recently, the formation of matter-wave soliton trains was experimentally reexamined

[33]. A nearly nondestructive imaging technique was used to follow the dynamics of these trains finding that a crucial role is played by noise and that neighboring solitons interact repulsively during the initial formation of the soliton train.

3.6 Problems

3.1. A gas of N classical particles is in equilibrium at temperature T in a three-dimensional isotropic harmonic trap with angular frequency ω. At time $t = 0$, the harmonic potential is abruptly removed. Determine the spatial density at later times. Repeat in the case of a zero-temperature Bose gas in the Thomas-Fermi approximation, and compare to the free expansion of the classical gas in the long time limit.

3.2. Derive the Bogoliubov spectrum starting from the Gross-Pitaevskii equation without performing the Madelung transformation.

3.3. Perform the dimensional reduction from 3D to 1D of the Gross-Pitaevskii action functional adopting the variational wave function:

$$\Psi(\mathbf{r}, t) = \frac{f(z, t)}{\pi^{1/2} \sigma a_\perp} \exp\left[-(x^2 + y^2)/(2\sigma^2 a_\perp^2)\right],$$

where σ is a dimensionless variational parameter and $a_\perp = [\hbar/(m\omega_\perp^2)]^{1/2}$, with $\omega_\perp$ the common angular frequency along x and y. Deduce the Euler-Lagrange equations for $f(z, t)$ and σ.

3.4. Calculate the superfluid fraction of a 3D interacting Bose gas under the assumption that the spectrum of the elementary excitations is phonon-like, i.e., $E(p) = c_s p$ with c_s the speed of sound.

3.5. Extending the procedure of Sect. 3.3.2, calculate the energy of two quantized vortices with opposite quantum number of circulation (vortex-antivortex pair).

3.6. Derive the formula for gray solitons, Eq. (3.39), assuming that $\phi(z - vt)$ of Eq. (3.36) is a complex wave function and using its modulus-phase decomposition.

3.7 Further Reading

- I. M. Khalatnikov, *An Introduction to the Theory of Superfluidity* (W. A. Benjamin, New York, 1965).
- D. R. Tilley and J. Tilley, *Superfluidity and superconductivity* (Taylor and Francis, 1990).

- A. Griffin, *Excitations in a Bose-Condensed Liquid* (Cambridge University Press, 1993).
- E. Timmermans, *Superfluids and superfluid mixtures in atoms traps*, Contemporary Physics **42**, 1 (2001).
- E. A. Cornell and C. E. Wieman, *Nobel Lecture: Bose-Einstein condensation in a dilute gas, the first 70 years and some recent experiments*, Rev. Mod. Phys. **74**, 875 (2002).
- W. Ketterle, *Nobel lecture: When atoms behave as waves: Bose-Einstein condensation and the atom laser*, Rev. Mod. Phys. **74**, 1131 (2002).
- A. Fetter, *Rotating trapped Bose-Einstein condensates*, Rev. Mod. Phys. **81**, 647 (2009).
- M. Tsubota, M. Kobayashi, and H. Takeuchi, *Quantum hydrodynamics*, Phys. Rep. **522**, 191 (2013).
- L. Pitaevskii and S. Stringari, *Bose-Einstein condensation and superfluidity* (Oxford University Press, 2016).
- T. Simula, *Quantized Vortices. A handbook of topological excitations* (Morgan and Claypool, San Rafael, 2019).
- B.V. Svistunov, E.S. Babaev, and N.V. Prokof'ef, *Superfluid States of Matter* (Taylor and Francis, Milton Park, 2021).

References

1. Abo-Shaeer, J.R., Raman, C., Vogels, J.M., Ketterle, W.: Observation of vortex lattices in Bose-Einstein condensates. Science **292**, 476 (2001)
2. Abo-Shaeer, J.R., Raman, C., Ketterle, W.: Formation and decay of vortex lattices in Bose-Einstein condensates at finite temperatures. Phys. Rev. Lett. **88**, 070409 (2002)
3. Anderson, M.H., Ensher, J.R., Matthews, M.R., Wieman, C.E., Cornell, E.A.: Observation of Bose-Einstein condensation in a dilute atomic vapor. Science **269**, 198 (1995)
4. Andrews, M.R., Townsend, C.G., Miesner, H.-J., Durfee, D.S., Kurn, D.M., Ketterle, W.: Direct, nondestructive observation of a Bose condensate. Science **273**, 84 (1996)
5. Andrews, M.R., Kurn, D.M., Miesner, H.-J., Durfee, D.S., Townsend, C.G., Inouye, S., Ketterle, W.: Propagation of sound in a Bose-Einstein condensate. Phys. Rev. Lett. **79**, 553 (1997)
6. Andrews, M.R., Townsend, C.G., Miesner, H.-J., Durfee, D.S., Kurn, D.M., Ketterle, W.: Observation of interference between two Bose condensates. Science **275**, 637 (1997)
7. Bogoliubov, N.N.: On the theory of superfluidity. J. Phys. (USSR) **11**, 23 (1947)
8. Bradley, C.C., Sackett, C.A., Tollett, J.J., Hulet, R.G.: Evidence of Bose-Einstein condensation in an atomic gas with attractive interactions. Phys. Rev. Lett. **75**, 1687 (1995) [Erratum, Phys. Rev. Lett. **79**, 1170 (1997)]
9. Burger, S., Bongs, K., Dettmer, S., Ertmer, W., Sengstock, K., Sanpera, A., Shlyapnikov, G.V., Lewenstein, M.: Dark solitons in Bose-Einstein condensates. Phys. Rev. Lett. **83**, 5198 (1999)
10. Chikkatur, A.P., Görlitz, A., Stamper-Kurn, D.M., Inouye, S., Gupta, S., Ketterle, W.: Suppression and enhancement of impurity scattering in a Bose-Einstein condensate. Phys. Rev. Lett. **85**, 483 (2000)
11. Chin, C., Bartenstein, M., Altmeyer, A., Riedl, S., Jochim, S., Hecker Denschlag, J., Grimm, R.: Observation of the pairing gap in a strongly interacting Fermi gas. Science **305**, 1128 (2004)
12. D'Agosta, R., Presilla, C.: States without a linear counterpart in Bose-Einstein condensates. Phys. Rev. A **65**, 043609 (2002)

13. Davis, K.B., Mewes, M.-O., Andrews, M.R., van Druten, N.J., Durfee, D.S., Kurn, D.M., Ketterle, W.: Bose-Einstein condensation in a gas of Sodium atoms. Phys. Rev. Lett. **75**, 3969 (1995)
14. DeMarco, B., Jin, D.S.: Onset of Fermi degeneracy in a trapped atomic gas. Science **285**, 1703 (1999)
15. Frisch, T., Pomeau, Y., Rica, S.: Transition to dissipation in a model of superflow. Phys. Rev. Lett. **69**, 1644 (1992)
16. Hadzibabic, Z., Gupta, S., Stan, C.A., Schunck, C.H., Zwierlein, M.W., Dieckmann, K., Ketterle, W.: Fiftyfold improvement in the number of quantum degenerate fermionic atoms. Phys. Rev. Lett. **91** 160401 (2003)
17. Henshaw, D.G., Woods, A.D.B.: Modes of atomic motions in liquid Helium by inelastic scattering of neutrons. Phys. Rev. **121**, 1266 (1961)
18. Ho, T.-L.: Spinor condensates in optical traps. Phys. Rev. Lett. **81**, 742 (1998)
19. Huepe, C., Brachet, M.-E.: Solutions de nucleation tourbillonnaires dans un modele d'ecoulement superfluide. C.R. Acad. Sci. Paris **325**, 195 (1997)
20. Inouye, S., Andrews, M.R., Stenger, J., Miesner, H.-J., Stamper-Kurn, D.M., Ketterle, W.: Observation of Feshbach resonances in a Bose-Einstein condensate. Nature **392**, 951 (1998)
21. Jackson, B., McCann, J.F., Adams, C.S.: Vortex formation in dilute inhomogeneous Bose-Einstein condensates. Phys. Rev. Lett **80**, 3903 (1998)
22. Keesom, W.H., Clusius, K.: On the anomaly in the specific heat of liquid Helium. Proc. Acad. Sci. Amst. **35**, 307, Common. Leider (1932)
23. Ketterle, W., van Druten, N.J.: Bose-Einstein condensation of a finite number of particles trapped in one or three dimensions. Phys. Rev. A **54**, 656 (1996)
24. Khaykovich, L., Schreck, F., Ferrari, G., Bourdel, T., Cubizolles, J., Carr, L.D., Castin, Y., Salomon, C.: Formation of a matter-wave bright soliton. Science **296**, 1290 (2002)
25. Kinast, J., Hemmer, S.L., Gehm, M.E., Turlapov, A., Thomas, J.E.: Evidence for Ssperfluidity in a resonantly interacting Fermi gas. Phys. Rev. Lett. **150**, 150402 (2004)
26. Kwon, W.J., Moon, G., Seo, S.W., Shin, Y.: Critical velocity for vortex shedding in a Bose-Einstein condensate. Phys. Rev. A **91**, 053615 (2015)
27. Lopes, R., Eigen, C., Navon, N., Clément, D., Smith, R.P., Hadzibabic, Z.: Quantum depletion of a homogeneous Bose-Einstein condensate. Phys. Rev. Lett. **119**, 190404 (2017)
28. Madison, K.W., Chevy, F., Wohlleben, W., Dalibard, J.: Vortex formation in a stirred Bose-Einstein condensate. Phys. Rev. Lett. **84**, 806 (2000)
29. Matthews, M.R., Anderson, B.P., Haljan, P.C., Hall, D.S., Wieman, C.E., Cornell, E.A.: Vortices in a Bose-Einstein condensate. Phys. Rev. Lett. **83**, 2498 (1999)
30. Mewes, M.O., Andrews, M.R., van Druten, N.J., Kurn, D.M., Durfee, D.S., Ketterle, W.: Bose-Einstein condensation in a tightly confining DC magnetic trap. Phys. Rev. Lett. **77**, 416 (1996)
31. Miesner, H.-J., Stamper-Kurn, D.M., Andrews, M.R., Durfee, D.S., Inouye, S., Ketterle, W.: Bosonic stimulation in the formation of a Bose-Einstein condensate. Science **279**, 1005 (1998)
32. Neely, T.W., Samson, E.C., Bradley, A.S., Davis, M.J., Anderson, B.P.: Observation of vortex dipoles in an oblate Bose-Einstein condensate. Phys. Rev. Lett. **104**, 160401 (2010)
33. Nguyen, J.H.V., Luo, D., Hulet, R.G.: Formation of matter-wave soliton trains by modulational instability. Science **356**, 422 (2017)
34. O'Hara, K.M., Hemmer, S.L., Gehm, M.E., Granade, S.R., Thomas, J.E.: Observation of a strongly interacting degenerate Fermi gas of atoms. Science **298**, 2179 (2002)
35. Onofrio, R., Raman, C., Vogels, J.M., Abo-Shaeer, J.R., Chikkatur, A.P., Ketterle, W.: Observation of superfluid flow in a Bose-Einstein condensed gas. Phys. Rev. Lett. **85**, 2228 (2000)
36. Raman, C., Köhl, M., Onofrio, R., Durfee, D.S., Kuklewicz, C.E., Hadzibabic, Z., Ketterle, W.: Evidence for a critical velocity in a Bose-Einstein condensed gas. Phys. Rev. Lett. **83**, 2502 (1999)
37. Raman, C., Abo-Shaeer, J.R., Vogels, J.M., Xu, K., Ketterle, W.: Vortex nucleation in a stirred Bose-Einstein condensate. Phys. Rev. Lett. **87**, 210402 (2001)

38. Regal, C.A., Greiner, M., Jin, D.S.: Observation of resonance condensation of fermionic atom pairs. Phys. Rev. Lett. **92**, 040403 (2004)
39. Stamper-Kurn, D.M., Andrews, M.R., Chikkatur, A.P., Inouye, S., Miesner, H.-J., Stenger, J., Ketterle, W.: Optical confinement of a Bose-Einstein condensate. Phys. Rev. Lett. **80**, 2027 (1998)
40. Stenger, J., Inouye, S., Andrews, M.R., Miesner, H.-J., Stamper-Kurn, D.M., Ketterle, W.: Strongly enhanced inelastic collisions in a Bose-Einstein condensate near Feshbach resonances. Phys. Rev. Lett. **82**, 2422 (1999)
41. Strecker, K.E., Partridge, G.B., Truscott, A.G., Hulet, R.G.: Formation and propagation of matter-wave soliton trains. Nature **417**, 150 (2002)
42. Truscott, A.G., Strecker, K.E., McAlexander, W.I., Partridge, G.B., Hulet, R.G.: Observation of Fermi pressure in a gas of trapped atoms. Science **291**, 2570 (2001)
43. Winiecki, T., McCann, J.F., Adams, C.S.: Pressure drag in linear and nonlinear quantum fluids. Phys. Rev. Lett. **82**, 5186 (1999)
44. Zwierlein, M.W., Stan, C.A., Schunck, C.H., Raupach, S.M.F., Kerman, A.J., Ketterle, W.: Condensation of pairs of fermionic atoms near a Feshbach resonance. Phys. Rev. Lett. **92**, 120403 (2004)

Ultracold Atoms as Strongly Correlated Systems

4

Ultracold atoms are obviously important in atomic physics and have played a major role in the improvement of high-resolution spectroscopy, as well as in the study of atomic scattering and the formation of ultracold molecules. However, it came as a surprise the range of applications and insights of relevance in condensed matter physics, with ultracold atoms playing the intermediate role between ideal, noninteracting systems fully dominated by quantum statistics and manifesting universal behavior and strongly correlated systems for which analytical treatments are lacking. From the theoretical standpoint, ultracold atoms are systems in which interactions can be controlled, offering therefore a continuum of possibilities precluded to the usual condensed matter systems. Due to their dilute character, the timescale and the microscopic lengthscale of many dynamical phenomena are longer and larger, respectively, and easily accessible using relatively simple optical imaging techniques. This research direction is ongoing, and it is both unpractical and premature to dare a discussion of all or many of the possible ramifications. Therefore, we will focus the attention on two aspects that, even historically, have been successfully studied with ultracold atoms in their strongly interacting regime: superfluidity properties of fermions and unconventional phase transitions. Even in their strongly interacting regime, ultracold atoms may manifest universal behavior useful to characterize quite different physical systems, such as superconductors and systems undergoing quantum phase transitions.

4.1 Interacting Fermi Systems

We have discussed interacting Bose systems in Chap. 1, leaving the description of their fermionic counterparts to this chapter, due to the many subtleties associated with interacting fermions. The key difference is the requirement for the complete antisymmetry of the many-body states which, in the presence of spin as necessary even for the simplest spin 1/2 fermions, imposes constraints on the possible arrangements and coupling. As it is manifest in their ground states for the noninteracting cases, see Fig. 1.1 for harmonic confinement, bosons and fermions have completely different behavior in a quantum degenerate regime. We expect this difference to persist also when we consider interacting particles, at least if the interactions are relatively weak. The simplest case to analyze, and also historically the first studied, is the case of an atom with many electrons, as first discussed by Llewellyn Thomas [29] and Enrico Fermi [6]. The solution of the many-body Schödinger equation for the Z electrons in an atom with $Z \gg 1$ is approximated by focusing, rather than on the many-body state, on the electron number density. This approach is the simplest of a class of models now called density functional theory (DFT). In this section, we explain some basic properties of interacting fermions in the spirit of the DFT, while the next section will be explicitly devoted to the Bardeen-Cooper-Schrieffer (BCS) model, a microscopic model of superconductivity, that requires a more sophisticated approach.

Having N fermions characterized by the two-body interaction potential $V(\mathbf{r}-\mathbf{r}')$ and under the effect of the external potential $U(\mathbf{r})$, one can consider their number density $n(\mathbf{r})$, normalized to N in the entire available volume. For the simplest situation of fermions with two spin components (spin 1/2) $\sigma = \uparrow, \downarrow$ one has $n(\mathbf{r}) = n_\uparrow(\mathbf{r}) + n_\downarrow(\mathbf{r})$, with $n_\uparrow(\mathbf{r}) = n_\downarrow(\mathbf{r}) = n(\mathbf{r})/2$ in a spin-balanced configuration.

In the Thomas-Fermi approach, the ground state of the many-body system is approximated imagining that each fermion is in equilibrium with all the others living at the same density $n(\mathbf{r})$, i.e., sharing the same spatially dependent Fermi energy. The kinetic energy term is then made by the integration of the energy over the entire volume, this term preventing the collapse of the electrons on the nucleus, enforcing the "Pauli principle":

$$T_{TF}[n] = \frac{3}{5}(3\pi^2)^{2/3}\frac{\hbar^2}{2m}\int d^3\mathbf{r}\, n(\mathbf{r})^{5/3}\,. \tag{4.1}$$

Obviously, this term is absent in the case of bosons that instead have a tendency to pile up in the same state even in the presence of mutual interactions, and this is the basis for what we called "Thomas-Fermi" approximation in Chap. 1.

This term for the total kinetic energy needs to be augmented by the energy for each electron due to the presence of a potential energy due to a given external field, such as the Coulomb field due to the nucleus in the case of an atom:

$$U[n] = \int d^3\mathbf{r}\, U(\mathbf{r})\, n(\mathbf{r})\,, \tag{4.2}$$

and the Coulomb repulsion between the electrons

$$E_D[n] = \frac{1}{2}\int d^3\mathbf{r}\, d^3\mathbf{r}'\, n(\mathbf{r})\, V(\mathbf{r}-\mathbf{r}')\, n(\mathbf{r}')\,, \tag{4.3}$$

4.1 Interacting Fermi Systems

In a variational approach, the ground state of this fermionic system can be obtained by considering the functional:

$$E_{TF}[n] = T_{TF}[n] + U[n] + E_D[n], \quad (4.4)$$

which should be minimized under the constraints $n(\mathbf{r}) \geq 0$ and

$$N = \int d^3\mathbf{r}\, n(\mathbf{r}). \quad (4.5)$$

This leads to an integral equation for the local fermionic density $n(\mathbf{r})$:

$$\frac{\hbar^2}{2m}(3\pi^2)^{2/3} n(\mathbf{r})^{2/3} + U(\mathbf{r}) + \frac{1}{2}\int d^3\mathbf{r} \int d^3\mathbf{r}'\, n(\mathbf{r})\, V(\mathbf{r}-\mathbf{r}')\, n(\mathbf{r}') = \mu, \quad (4.6)$$

with μ the Lagrange multiplier fixed by the normalization in Eq. (4.5), which can be also identified as the Fermi energy of the interacting system.

Equation (4.6) can be solved numerically by using an iterative procedure, choosing an initial guess for the density, calculating the terms in its left-hand side, then adjusting the density for self-consistency, and repeating the process until a targeted precision is achieved. This is equivalent to search for the minimization of the functional Eq. (4.4) by sampling over a large number of densities $n(\mathbf{r})$.

The Thomas-Fermi density functional for spin 1/2 fermions uniformly confined in a cubic box of size L and in the presence of a interparticle contact interaction of strength g is written as

$$E[n_\uparrow, n_\downarrow] = \int_{L^3} d^3\mathbf{r} \left\{ \frac{3}{5}\frac{\hbar^2}{2m}(6\pi^2)^{2/3}\left(n_\uparrow^{5/3} + n_\downarrow^{5/3}\right) + g n_\uparrow n_\downarrow \right\}. \quad (4.7)$$

The minimization of the energy functional with the constraint of fixed total number of fermions gives

$$\frac{\hbar^2}{2m}(6\pi^2)^{2/3} n_\uparrow^{2/3} + g n_\downarrow = \mu_\uparrow, \quad (4.8)$$

$$\frac{\hbar^2}{2m}(6\pi^2)^{2/3} n_\downarrow^{2/3} + g n_\uparrow = \mu_\downarrow, \quad (4.9)$$

with $\mu_\uparrow$ and $\mu_\downarrow$ the Lagrange multipliers of the constrained minimization, the chemical potentials for each spin species. This system is energetically stable only if the Hessian

$$\begin{pmatrix} \frac{\partial E}{\partial n_\uparrow^2} & \frac{\partial E}{\partial n_\uparrow \partial n_\downarrow} \\ \frac{\partial E}{\partial n_\downarrow \partial n_\uparrow} & \frac{\partial E}{\partial n_\downarrow^2} \end{pmatrix} = L^3 \begin{pmatrix} \frac{2}{3}\frac{\hbar^2}{2m}(6\pi^2)^{2/3} n_\uparrow^{-1/3} & g \\ g & \frac{2}{3}\frac{\hbar^2}{2m}(6\pi^2)^{2/3} n_\downarrow^{-1/3} \end{pmatrix} \quad (4.10)$$

is positive defined. Thus, the uniform and unpolarized configuration, $n_\uparrow = n_\downarrow = n/2$, is stable only if

$$0 \leq g \leq \frac{2}{3}(6\pi^2)^{2/3} \frac{\hbar^2}{2m} \left(\frac{n}{2}\right)^{-1/3}. \tag{4.11}$$

If g exceeds the expression on the last term of this inequality, an instability arises, called Stoner instability [28], invoked to describe, within a mean-field approach, the emergence of a ferromagnetic phase in an electron gas with strong Coulomb repulsion between oppositely oriented electron spins. This instability has been observed experimentally in an ultracold mixture of fermionic atoms at LENS [31]. For $g < 0$, attractive interaction between fermions with opposite spin, there is another instability, named after Cooper. This instability is an artifact of the Thomas-Fermi model, however signaling that attractive fermions need a less simplified theoretical description.

It is important to stress that the Thomas-Fermi density functional does not include the exchange energy term. However, Eq. (4.6) can be used as a simple theoretical tool to obtain the ground-state density profile $n(\mathbf{r})$, and the corresponding ground-state energy, of an interacting gas of N identical fermions. Within the DFT framework, a more accurate description is obtained by using another density functional, proposed by Walter Kohn and Lu Jeu Sham [Kohn (1965)], where N single-particle orbitals $\phi_i(\mathbf{r})$ determine the local density, namely,

$$n(\mathbf{r}) = \sum_{i=1}^{N} |\phi_i(\mathbf{r})|^2, \tag{4.12}$$

and the expression for the kinetic energy functional

$$T_{KS}[n] = \sum_{i=1}^{N} \int d^3\mathbf{r}\, \phi_i^*(\mathbf{r}) \left(-\frac{\hbar^2}{2m}\right) \nabla^2 \phi_i(\mathbf{r}), \tag{4.13}$$

which is superseding Eq. (4.1). In addition, the Kohn-Sham functional energy contains the exchange energy omitted in the Thomas-Fermi approach. Nowadays, the Kohn-Sham DFT is extremely useful for an accurate description of electrons in atoms and molecules. However, the Kohn-Sham approach requires the solution of N coupled nonlinear Schrödinger equations for the single-particle orbitals $\phi_i(\mathbf{r})$. This makes the Kohn-Sham method impracticable in the case of 10^3–10^6 atoms, as usual for trapped ultracold Fermi gases.

In the condensed matter physics setting, the typical situation involves Avogadro number of particles, and specific techniques in this many-body limit have been developed, in particular the Landau model of Fermi liquid. This model assumes that the Fermi-Dirac distribution function is not drastically altered by the presence of weak interparticle interactions and that the case of interacting fermions can be obtained "adiabatically" or "continuously" (depending on assuming a physical or mathematical standpoint) from the noninteracting case. Under this assumption, the effect of the interaction is to renormalize some parameters of the fermions, such as the mass. As in the case of interacting bosons discussed in Chap. 3, with the Bogoliubov phonons as collective excitations, we can replace interacting fermions with noninteracting quasi-fermions obtained from the former via renormalized parameters and a modified dispersion relationship.

This phenomenological approach has been applied to describe in the quantum degenerate regime both electrons in a crystal lattice and liquid ^{3}He and has been also corroborated by more microscopic approaches. For instance, electron-electron interactions are mediated in vacuum by photon exchange, the direct Coulomb interaction. In a crystal, there is also the possibility for the electron to polarize the surrounding atoms. Therefore, an electron in solid state will not move as a free electron in vacuum but will instantaneously bring with it a surrounding cloud of polarized atoms, forming a quasiparticle. This quasiparticle, called polaron, has the same electric charge of the electron but a different mass, typically larger in metals, smaller in semiconductors. A first important consequence is the existence of a mechanism to release energy from electrons to the ion lattice. For instance, if electrons are accelerated by an external electric field, they can transfer energy to the lattice via phonon emission. This is the mechanism giving rise to a nonzero resistivity even assuming an ideal lattice, without impurities, lattice defects, and surface contributions. Moreover, in this way, the direct Coulomb interaction between two electrons is screened by the ion lattice, as understandable by the following dynamics. Suppose an electron is suddenly added in the crystal, then the surrounding electrons will be repelled, and the lattice ions will be attracted. This propagates a "charge wave," a perturbation of the local charge density, even at long distance from the created electron. Such a collective excitation is described in terms quasiparticles named plasmons. Then, electrons in an ideal lattice, as far as their interaction with the lattice is concerned, can be replaced by corresponding quasiparticles, the polarons, interacting with collective excitations in the form of extended density perturbations of the lattice, the phonons, and localized perturbations of the charge density, plasmons. As we will discuss in the next section, this leads to the surprising possibility of attractive electron-electron interactions for highly selective states of electron pairs.

4.2 BCS Model

Fermions may both couple together to form effective bosons—after all as commented earlier all bosonic atoms are made of fermions—and under specific circumstances, they also couple in a peculiar many-body fashion. In solid-state physics, the most relevant fermions are the conduction electrons, which can quasi-freely move inside the material. To "bosonize," a system of electrons should be arranged in a bound state with total angular momentum, including the spin contribution, forming an integer number in units of $\hbar$. This is, for instance, achieved by two electrons bound in an s-wave state of angular momentum and a singlet spin state. However, a mechanism to create an effective attraction between the two electrons is necessary for them to bind, as in vacuum they just experience the bare Coulomb repulsion. The discussion of the former section is instrumental to this goal, as the presence of collective excitations in the lattice can upset the electron-electron interaction to create binding.

In 1957, John Bardeen, Leon Cooper, and Robert Schrieffer suggested that the superconductivity observed at sufficiently low temperature in various materials is due to the formation of electron pairs in singlet spin states [2]. This happens only for electrons with energy near the Fermi surface within the Debye energy of the lattice. This can be argued by considering a second-quantized approach to the problem (see also Appendix A). In the simplest case of electrons coupled via phonons, the interaction Hamiltonian operator can be written as

$$\hat{H}_{\text{el-ph}} = \sum_{\mathbf{k},\mathbf{q},\sigma} M_{\mathbf{q}} (\hat{a}_{\mathbf{q}} + \hat{a}^+_{-\mathbf{q}}) \hat{c}^+_{\mathbf{k}+\mathbf{q},\sigma} \hat{c}_{\mathbf{k},\sigma}, \qquad (4.14)$$

where we have introduced annihilation and creation operators for phonons with wave vectors $\mathbf{q}$ and annihilation and creation operators for electrons with wave vectors $\mathbf{k}$. The first term in the right-hand side of Eq. (4.14) corresponds to a vertex diagram in which an electron of wave vector $\mathbf{k}$ and a phonon with wave vector $\mathbf{q}$ are destroyed to create electron with wave vector $\mathbf{k}+\mathbf{q}$. The second represents a vertex diagram in which an electron of wave vector $\mathbf{k}$ is destroyed to create an electron with wave vector $\mathbf{k}+\mathbf{q}$ and a phonon of wave vector $-\mathbf{q}$. Then, both processes of emission and absorption of phonons are covered, with conservation of energy and momentum, and no change in the spin state of the electrons. The quantity $M_{\mathbf{q}}$ represents the amplitude of the process, depending on the strength of the electron-phonon interaction. Incidentally, the relevance of the electron-phonon interaction to explain superconductivity is shown by the fact that good conductors, such as copper and silver, do not become superconductors.

The vertices discussed above can be combined to produce an effective electron-electron interaction mediated by phonons. This can be represented in diagrammatic terms and, under the usual conditions of applicability of perturbation theory, at the leading order, there are two contributions with absorption or emission of a phonon of wave vector $\mathbf{q}$ or $-\mathbf{q}$ (see Fig. 4.1). For the corresponding amplitude between an initial state $|i\rangle$ and a final state $|f\rangle$ only made of electrons, with the phonons just present as intermediate states, the scattering amplitude will be proportional to the phonon-electron interaction coupling strength $M_{\mathbf{q}}$ appearing in Eq. (4.14), times the matrix element

$$\frac{\langle f | \hat{c}^+_{\mathbf{k}+\mathbf{q},\sigma} \hat{c}_{\mathbf{k},\sigma} \hat{a}_{\mathbf{q}} | m \rangle \langle m | \hat{c}^+_{\mathbf{k}'-\mathbf{q},\sigma'} \hat{c}_{\mathbf{k}',\sigma'} \hat{a}^+_{\mathbf{q}} | i \rangle}{E(\mathbf{k}') - E(\mathbf{k}'-\mathbf{q}) - \hbar \omega_{\mathbf{q}}}$$

$$+ \frac{\langle f | \hat{c}^+_{\mathbf{k}'-\mathbf{q},\sigma'} \hat{c}_{\mathbf{k}',\sigma'} \hat{a}_{-\mathbf{q}} | m' \rangle \langle m' | \hat{c}^+_{\mathbf{k}+\mathbf{q},\sigma} \hat{c}_{\mathbf{k},\sigma} \hat{a}^+_{-\mathbf{q}} | i \rangle}{E(\mathbf{k}) - E(\mathbf{k}+\mathbf{q}) + \hbar \omega_{\mathbf{q}}}, \qquad (4.15)$$

4.2 BCS Model

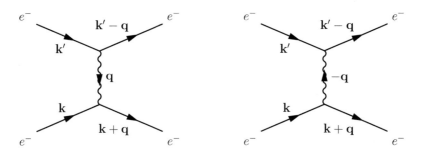

Fig. 4.1 Diagrams for the contribution to the lowest order in perturbation theory to electron-electron scattering mediated by phonons, corresponding to the first term (left graph) and the second term (right graph) in Eq. (4.15)

which is summed over intermediate states $|m\rangle$, $|m'\rangle$ with all possible values of $\mathbf{k}, \mathbf{k'}, \mathbf{q}, \sigma, \sigma'$. Notice, for instance, looking at the energy conservation in the left diagram in Fig. 4.1, that $E(\mathbf{k'}) = E(\mathbf{k'} - \mathbf{q}) + \hbar\omega_\mathbf{q}$, $E(\mathbf{k}) + \hbar\omega_\mathbf{q} = E(\mathbf{k} + \mathbf{q})$.

The matrix element in Eq. (4.15) can be summed over the intermediate states for the phonons finally yielding a scattering amplitude containing an effective Hamiltonian with matrix elements:

$$\langle f|\hat{H}_{\text{eff}}|i\rangle = \frac{1}{2} \sum_{\mathbf{k}\mathbf{k'}\mathbf{q},\sigma,\sigma'} V_{\mathbf{k}\mathbf{q}} \langle f|\hat{c}^+_{\mathbf{k}+\mathbf{q},\sigma}\hat{c}^+_{\mathbf{k'}-\mathbf{q},\sigma'}\hat{c}_{\mathbf{k},\sigma}\hat{c}_{\mathbf{k'},\sigma'}|i\rangle \quad (4.16)$$

where

$$V_{\mathbf{k}\mathbf{q}} = |M_\mathbf{q}|^2 \left[\frac{1}{E(\mathbf{k}) - E((\mathbf{k}+\mathbf{q})) - \hbar\omega_\mathbf{q}} - \frac{1}{E(\mathbf{k}) - E((\mathbf{k}+\mathbf{q})) + \hbar\omega_\mathbf{q}} \right]$$

$$= 2|M_\mathbf{q}|^2 \frac{\hbar\omega_\mathbf{q}}{[E(\mathbf{k}+\mathbf{q}) - E(\mathbf{k})]^2 - (\hbar\omega_\mathbf{q})^2}. \quad (4.17)$$

This last expression shows that the sign of $V_{\mathbf{k}\mathbf{q}}$ depends on the sign of the denominator in the last expression on the right-hand side, $[E(\mathbf{k}+\mathbf{q}) - E(\mathbf{k})]^2 - (\hbar\omega_\mathbf{q})^2$. If the scattering involves two electrons with momenta $\mathbf{k}$ differing by less than $\omega_\mathbf{q}$, the sign is negative, and then, the electron-electron interaction becomes attractive. Since $\omega_\mathbf{q}$ has an upper bound, the propagation of phonons corresponding to a wavelength smaller than the lattice size is not allowed; this makes a narrow set of electron pairs which can bind together via a strong correlation in momentum space, the so-called Cooper pairs. This strong correlation in momentum space corresponds to a loose correlation in ordinary space, so we expect the Cooper pair to have a spatial wave function extended over large distance with respect to the lattice spacing and the average distance between any two free electrons. Thus, these Cooper pairs are intrinsically many-body states, with no possibility to conceive them in their individuality. Due to the impossibility of electronic scattering in states with full occupation, well below the Fermi energy, the only electrons available for pairing are the ones on the Fermi surface, within an energy distance from it smaller than the Debye energy.

Fortunately, in systems like ^{3}He and ultracold fermions, one does not have to invoke such a subtle mechanism for pairing, since the direct interatomic potential can make it. We present now a general approach, based on second quantization, for a generic interatomic potential.

The Hamiltonian operator of a system of nonrelativistic interacting two-spin-component fermions is given by

$$\hat{H} = \sum_{\sigma=\uparrow,\downarrow} \int d^3\mathbf{r}\, \hat{\psi}_\sigma^+(\mathbf{r}) \left(-\frac{\hbar^2}{2m}\nabla^2 + U_\sigma(\mathbf{r}) \right) \hat{\psi}_\sigma(\mathbf{r})$$
$$+ \frac{1}{2} \sum_{\sigma,\sigma'=\uparrow,\downarrow} \int d^3\mathbf{r}\, d^3\mathbf{r}'\, \hat{\psi}_\sigma^+(\mathbf{r})\, \hat{\psi}_{\sigma'}^+(\mathbf{r}')\, V_{\sigma\sigma'}(\mathbf{r}-\mathbf{r}')\, \hat{\psi}_{\sigma'}(\mathbf{r}')\, \hat{\psi}_\sigma(\mathbf{r}),$$

(4.18)

where $\hat{\psi}_\sigma(\mathbf{r})$ is the fermionic field operator, $U_\sigma(\mathbf{r})$ the external potential acting on fermions with spin σ, and $V_{\sigma\sigma'}(\mathbf{r}-\mathbf{r}')$ the interparticle interaction.

In this case, the many-body ground state $|GS\rangle$ of the system is quite different with respect to the one of noninteracting fermions. The Cooper pairs are described by the pairing fields:

$$\Delta_{\uparrow\downarrow}(\mathbf{r}) = -\int d^3\mathbf{r}'\, V_{\uparrow\downarrow}(\mathbf{r}-\mathbf{r}')\, \langle \hat{\psi}_\uparrow(\mathbf{r})\hat{\psi}_\downarrow(\mathbf{r}') \rangle, \quad (4.19)$$

$$\Delta_{\downarrow\uparrow}(\mathbf{r}) = -\int d^3\mathbf{r}'\, V_{\downarrow\uparrow}(\mathbf{r}-\mathbf{r}')\, \langle \hat{\psi}_\downarrow(\mathbf{r})\hat{\psi}_\uparrow(\mathbf{r}') \rangle, \quad (4.20)$$

where $\langle \ldots \rangle$ indicates the ground-state average $\langle GS|\ldots|GS\rangle$. One can prove that the pairing fields are zero only for fermions having a purely repulsive interaction. This is a key point: only fermions with a generic interaction potential $V_{\sigma\sigma'}(\mathbf{r}-\mathbf{r}')$ which contains an attractive part have a finite pairing. In the simplest cases, the interparticle interaction does not depend on the spin, i.e., $V_{\uparrow\downarrow}(\mathbf{r}-\mathbf{r}') = V(\mathbf{r}-\mathbf{r}')$, and we have only one pairing field $\Delta(\mathbf{r})$. Moreover, let us consider a simple model of the attractive fermion-fermion interaction with a pointlike potential, as we did also in the case of interacting bosons (the Fermi pseudopotential in Eq. (1.117)):

$$V(\mathbf{r}-\mathbf{r}') = g\, \delta(\mathbf{r}-\mathbf{r}'), \quad (4.21)$$

with $g < 0$ for attractive interactions. Consequently, we have

$$\Delta(\mathbf{r}) = -g\, \langle \hat{\psi}_\uparrow(\mathbf{r})\hat{\psi}_\downarrow(\mathbf{r}) \rangle = -g\, \langle \hat{\psi}_\downarrow(\mathbf{r})\hat{\psi}_\uparrow(\mathbf{r}) \rangle. \quad (4.22)$$

4.2 BCS Model

In a uniform configuration where $U_\sigma(\mathbf{r}) = 0$, it is a reasonable approximation to assume that the pairing field Δ does not depend on space.

Even assuming the simplest case of a uniform configuration and a contact interaction only for fermions with opposite spins, the many-body problem cannot be solved analytically due to the presence of the quartic term $\hat{\psi}_\uparrow^+ \hat{\psi}_\downarrow^+ \hat{\psi}_\downarrow \hat{\psi}_\uparrow$ in the simplified Hamiltonian:

$$\hat{H} = \int_{L^3} d^3\mathbf{r} \left\{ -\frac{\hbar^2}{2m} \sum_{\sigma=\uparrow,\downarrow} \hat{\psi}_\sigma^+ \nabla^2 \hat{\psi}_\sigma + g\, \hat{\psi}_\uparrow^+ \hat{\psi}_\downarrow^+ \hat{\psi}_\downarrow \hat{\psi}_\uparrow \right\}. \quad (4.23)$$

Bardeen, Cooper, and Schrieffer proposed a decoupling approximation (see Problem 4.5), assumed valid whenever $g < 0$ and there is sizable pairing Δ, which is at the basis of their BCS model:

$$g\, \hat{\psi}_\uparrow^+ \hat{\psi}_\downarrow^+ \hat{\psi}_\downarrow \hat{\psi}_\uparrow \simeq -\Delta \hat{\psi}_\downarrow \hat{\psi}_\uparrow - \Delta \hat{\psi}_\uparrow^+ \hat{\psi}_\downarrow^+ - \frac{\Delta^2}{g}, \quad (4.24)$$

In this way, the Hamiltonian (4.23) becomes quadratic:

$$\hat{H}_{BCS} = \int_{L^3} d^3\mathbf{r} \left\{ -\frac{\hbar^2}{2m} \sum_{\sigma=\uparrow,\downarrow} \hat{\psi}_\sigma^+ \nabla^2 \hat{\psi}_\sigma - \Delta \hat{\psi}_\downarrow \hat{\psi}_\uparrow - \Delta \hat{\psi}_\uparrow^+ \hat{\psi}_\downarrow^+ - \frac{\Delta^2}{g} \right\}. \quad (4.25)$$

This is the BCS Hamiltonian used to investigate the properties of fermions with attractive interaction. Due to the fact that this Hamiltonian is now quadratic with respect to the fermionic operators, one can diagonalize the shifted Hamiltonian $\hat{H}_{BCS} - \mu \hat{N}$, with μ the chemical potential of the system. This "rotation" is obtained by means of the Bogoliubov-Valatin transformation [3, 30]:

$$\hat{\psi}_\sigma(\mathbf{r}) = \frac{1}{L^{3/2}} \sum_{\mathbf{k}} \left(\hat{b}_{\mathbf{k}\sigma}\, u_q\, e^{i\mathbf{k}\cdot\mathbf{r}} + \hat{b}_{\mathbf{k}\sigma'}^+\, v_q\, e^{-i\mathbf{k}\cdot\mathbf{r}} \right) \quad (4.26)$$

with $\sigma' \neq \sigma$, where u_k and v_k are the so-called Bogoliubov amplitudes (satisfying $|u_k|^2 + |v_k|^2 = 1$), L is the size of the cubic volume L^3 containing the system, and $\mathbf{q}$ is the wave vector of the two-particle scattering problem. After imposing anticommutation rules to the Bogoliubov ladder operators $\hat{b}_{\mathbf{k}\sigma}$ and $\hat{b}_{\mathbf{k}\sigma}^+$, and some algebraic manipulations, we obtain, apart from a relative phase factor, the amplitudes:

$$u_k = \frac{1}{2}\left[1 - \frac{1}{E_\mathbf{k}}\left(\frac{\hbar^2 k^2}{2m} - \mu\right)\right], \quad v_k = \frac{1}{2}\left[1 + \frac{1}{E_\mathbf{k}}\left(\frac{\hbar^2 k^2}{2m} - \mu\right)\right]. \quad (4.27)$$

which allow to write the diagonalized Hamiltonian as

$$\hat{H}_{BCS} - \mu \hat{N} = -\frac{\Delta^2}{g} L^3 + \sum_{\mathbf{k}} \left(\frac{\hbar^2 k^2}{2m} - \mu - E_k \right) + \sum_{\sigma=\uparrow\downarrow} \sum_{\mathbf{k}} E_{\mathbf{k}} \hat{b}^+_{\mathbf{k}\sigma} \hat{b}_{\mathbf{k}\sigma}, \quad (4.28)$$

provided that the dispersion relationship of the resulting fermionic quasiparticles is given by

$$E_k = \sqrt{\left(\frac{\hbar^2 k^2}{2m} - \mu \right)^2 + \Delta^2} . \quad (4.29)$$

By inspection, one recognizes that Δ plays the role of an energy gap. The presence of this finite energy gap Δ is a sufficient condition for the superfluid behavior of the fermionic system, because excess energy provided from, for instance, an external electric field cannot be released unless its amount is larger than the energy gap. By performing the average with respect to the ground state $|GS\rangle$ of the shifted diagonal Hamiltonian (4.28), we have

$$\langle \hat{H}_{BCS} - \mu \hat{N} \rangle = -\frac{\Delta^2}{g} L^3 + \sum_{\mathbf{k}} \left(\frac{\hbar^2 k^2}{2m} - \mu - E_q \right) . \quad (4.30)$$

The energy gap Δ due to pairing can be determined by extremizing this energy:

$$\frac{\partial}{\partial \Delta} \langle \hat{H}_{BCS} - \mu \hat{N} \rangle = 0 . \quad (4.31)$$

In this way, we obtain the so-called $T - 0$ gap equation:

$$-\frac{1}{g} = \frac{1}{2L^3} \sum_{\mathbf{k}} \frac{1}{E_k(\Delta)} . \quad (4.32)$$

The gap energy Δ can be numerically evaluated by fixing the chemical potential μ and the interaction strength g. However, to avoid a quadratic divergence, since the terms in the sum of Eq. (4.32) scales as q^{-1}, one must also introduce an ultraviolet cutoff Λ such that $|\mathbf{k}| < \Lambda$.

In the case of ultracold fermions, there seems to be no immediate analog of the weak attraction between Cooper pairs as it happens in superconductors. It is possible to mimic this interaction by including a Bose gas as a medium, realizing a properly prepared Bose-Fermi mixture in an optical lattice [10]; however, atoms have the advantage that they do not have a dominant Coulomb repulsion to be overcome for pairing and that the sign of the interatomic interaction can be attractive or repulsive. In the latter case, we do not expect the onset of a BCS regime, but fermions can form effective bosons in the form of tight molecular dimers, with a strong correlation in

4.2 BCS Model

ordinary space. These pairs are also named Schafroth pairs, by a model suggested by Schafroth predating the BCS model [25, 26]. In 1969, David Eagles suggested the possibility of having, for a fermionic superconductor, a crossover from weakly bound BCS-like Cooper pairs to the Bose-Einstein condensation (BEC) of strongly bound dimers [5]. In 1980, Anthony Leggett showed that, in the context of a dilute ultracold Fermi gas, scattering theory plays an essential role in the description of this BCS-BEC crossover [16]. One of the distinctive features of superfluidity of strongly bound dimers is that the critical temperature for superfluidity coincides with the critical temperature for Bose-Einstein condensation of the bosonized pair. Although the critical temperatures for the superconducting materials first investigated were measured in the same range of the critical temperature for the superfluidity of ^{4}He, it turns out that this is purely accidental, as we will discuss when taking into full account the effect of finite temperatures.

In Chap. 1, we showed that in the case of a contact interaction potential between two particles of mass m the first-order Born approximation relates the interaction strength g to the s-wave scattering length a by Eq. (1.120), repeated here for convenience

$$g = \frac{4\pi\hbar^2}{m} a . \tag{4.33}$$

However, a calculation carried out at second order in the Born approximation gives a more involved relationship between g and a, namely,

$$\frac{m}{4\pi\hbar^2 a} = \frac{1}{g} + \frac{1}{L^3} \sum_{\mathbf{k}} \frac{m}{\hbar^2 k^2} , \tag{4.34}$$

This equation may be also derived from the Lippmann-Schwinger equation for the T matrix (see Appendix C). To avoid the linear divergence of the second term on the right side, we limit the sum to $|\mathbf{k}| < \Lambda$, with Λ an ultraviolet cutoff.

Since g can assume both signs while the last term in Eq. (4.34) is positive defined, a can also change sign, but unlike g in doing so it diverges. In the continuum limit $\sum_{\mathbf{k}} \to L^3 \int d^3\mathbf{k}/(2\pi)^3$, after integration over momenta, it reads

$$\frac{m}{4\pi\hbar^2 a} = \frac{1}{g} + \frac{m}{2\pi^2\hbar^2} \Lambda . \tag{4.35}$$

A possible interpretation of this result is that $4\pi\hbar^2 a/m$ is the "regularized" interaction strength g_R, expressed in terms of the "bare" interaction strength g by the relation $g_R^{-1} = g^{-1} + m\Lambda/(2\pi^2\hbar^2)$.

Three distinct regimes can be identified based on Eq. (4.35). If $g \to 0^\pm$, for any finite Λ, the scattering length a will also follow the same limit approaching zero with the same sign, $a \to 0^\pm$. This implies that if the scattering length can be made small and positive, formation of dimers is possible, and superfluidity will be related to their nature as effective Bose-Einstein condensates. If instead the scattering length is small and negative, BCS pairing can occur. A third situation

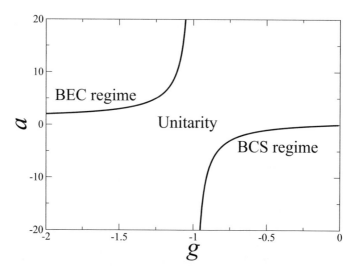

Fig. 4.2 Fermionic scattering length a as a function of the bare interaction strength g, for a finite ultraviolet cutoff Λ; see Eq. (4.35). In the plot, a is in units of $m/(4\pi\hbar^2)$ and $m\Lambda/(2\pi^2\hbar^2) = 1$

occurs when $g \to -m\Lambda/(2\pi^2\hbar^2)^{\pm}$, since a will diverge, $a \to \mp\infty$, as shown in Fig. 4.2. This regime of large scattering length is called unitarity regime, as the interparticle energy dominates any other aspect of the dynamics, only limited by the general request for unitarity in scattering theory. In spite of the divergence of the scattering length and the strong coupling regime for g_R, we will see that the superfluid properties of fermions in this intermediate regime do not have dramatic behavior and that relevant indicators are continuous functions of the available "knobs," a uniform external magnetic field from the experimental standpoint. For this reason, this regime is also named BCS-BEC (or BEC-BCS) "crossover."

A further feature of Eq. (4.35) is that a bound state is possible only if $g < -2\pi^2\hbar^2/(m\Lambda)$. In the presence of a bound state of energy $-\epsilon_B$, it follows

$$-\frac{1}{g} = \frac{1}{L^3} \sum_{|\mathbf{k}|<\Lambda} \frac{m}{\hbar^2 k^2 + m\epsilon_B} \ . \qquad (4.36)$$

After integration over momenta, one obtains

$$-\frac{1}{g} = \frac{m}{2\pi^2\hbar^2}\left[\Lambda - \sqrt{m\epsilon_B/\hbar^2}\arctan\left(\Lambda/\sqrt{m\epsilon_B/\hbar^2}\right)\right] \ , \qquad (4.37)$$

which implies, by comparing Eqs. (4.35) and (4.37)

$$a = \sqrt{\frac{\hbar^2}{m\epsilon_B}}\frac{\pi}{2}\arctan^{-1}\left(\Lambda/\sqrt{m\epsilon_B/\hbar^2}\right) \ . \qquad (4.38)$$

4.2 BCS Model

In the limit $\Lambda \to +\infty$, one gets a remarkably simple expression for the relationship between the energy of the bound state and the scattering length a, $\epsilon_B = \hbar^2/(ma^2)$.

4.2.1 BCS Pairing at Finite Temperature

In this subsection, we show how the BCS-BEC crossover can be theoretically investigated by using Eq. (4.34) and its consequences with particular focus on the influence of a finite temperature. We consider a system of nonrelativistic interacting spin 1/2 fermions at finite temperature T and fixed chemical potential μ. This choice suggests the use of the grand canonical ensemble, taking into account the BEC Hamiltonian of Eq. (4.25). The grand canonical partition function and the grand canonical potential of the system are respectively expressed by Eqs. (1.29) and (1.33) with $\hat{H} = \hat{H}_{BCS}$ and $\hat{N} = \hat{N}_\uparrow + \hat{N}_\downarrow$. The number operators for each spin component $\sigma = \uparrow, \downarrow$ can also be expressed in terms of the corresponding field operators as

$$\hat{N}_\sigma = \int_{L^3} d^3\mathbf{r}\, \hat{n}_\sigma(\mathbf{r}) = \int_{L^3} d^3\mathbf{r}\, \hat{\psi}_\sigma^+(\mathbf{r})\hat{\psi}_\sigma(\mathbf{r}) , \qquad (4.39)$$

the number operator for fermions with spin σ and $\hat{n}_\sigma(\mathbf{r})$ the local density operator. The thermal average of number operators reads

$$\langle \hat{N}_\sigma \rangle = \frac{1}{\mathcal{Z}} Tr[\hat{N}_\sigma\, e^{-\beta(\hat{H}_{BCS} - \mu\hat{N})}] . \qquad (4.40)$$

We suppose that the system is in a state with $n_\uparrow = n_\downarrow = n/2$, with $n = N/V$ the total number density of fermions and $V = L^3$ the confinement volume in a cubic box of size L.

After some mathematical manipulations and considering that BCS dispersion relationship in Eq. (4.29), one finds that the grand canonical potential can be written as

$$\Omega = \Omega_0 + \Omega_T , \qquad (4.41)$$

where

$$\Omega_0 = -\frac{\Delta^2}{g} L^3 + \sum_{\sigma=\uparrow\downarrow} \sum_{\mathbf{k}} \left(\frac{\hbar^2 k^2}{2m} - \mu - E_k \right) , \qquad (4.42)$$

is the contribution which reduced to the ground-state one at zero temperature as from Eq. (4.30), while

$$\Omega_T = \frac{2}{\beta} \sum_{\mathbf{k}} \ln(1 + e^{-\beta E_k}) , \qquad (4.43)$$

is the contribution of the fermionic single-particle excitations existing at nonzero temperature. Notice that Δ in Eq. (4.42) is temperature dependent and that also appears in Eq. (4.43) in the expression for the energies E_k of the fermionic excitations.

To determine the energy gap Δ which appears in the fermionic energy spectrum (4.29), we impose minimization of the grand canonical potential:

$$\frac{\partial \Omega}{\partial \Delta} = 0 \,. \tag{4.44}$$

Thus, we are finding the pairing Δ which minimizes the grand potential Ω, obtaining

$$-\frac{1}{g} = \frac{1}{L^3} \sum_{\mathbf{k}} \frac{1}{2E_k} \tanh\left(\frac{\beta E_k}{2}\right) \,. \tag{4.45}$$

This formula reduces, in the $\beta \to +\infty$ limit, to Eq. (4.32) derived in the previous section. The average number density n is instead derived by using the third expression in Eq. (1.36) with $n = N/V = N/L^3$, obtaining

$$n = \frac{1}{L^3} \sum_{\mathbf{k}} \left(1 - \frac{\hbar^2 k^2 - \mu}{2m E_k}\right) \tanh\left(\frac{\beta E_k}{2}\right) \,. \tag{4.46}$$

The gap equation (4.45) is ultraviolet divergent. As we did previously with the formula (4.34) for the scattering length, by introducing the ultraviolet cutoff Λ, we get a convergent relationship:

$$-\frac{1}{g} = \frac{1}{L^3} \sum_{|\mathbf{k}|<\Lambda} \frac{1}{E_k} \tanh\left(\frac{\beta E_k}{2}\right) \,. \tag{4.47}$$

Deep BCS Regime for Superconductors

For electrons in superconducting metals, the cutoff energy $\hbar^2 \Lambda^2/(2m)$ can be identified with the Debye energy $\hbar \omega_D$, where ω_D is the Debye frequency, that is the maximum frequency of vibration for the atoms that make up the ionic crystal of the metal. Moreover, the chemical potential μ can be safely approximated also at finite temperature by the Fermi energy of noninteracting fermions. This approximation, fully justified for electrons in metals, identifies the so-called deep BCS regime of the BCS-BEC crossover.

In the thermodynamic limit, where the volume L^3 goes to infinity, $\sum_{\mathbf{k}}$ can be replaced by $L^3 \int d^3\mathbf{k}/(2\pi)^3 = L^3 \int N(\xi) d\xi$ with $N(\xi) = \int d^3\mathbf{k}/(2\pi)^3 \, \delta(\xi - \xi_k)$, $\xi_k = \epsilon_k - \epsilon_F$, and $\epsilon_k = \hbar^2 k^2/(2m)$. In metals, the condition $\hbar \omega_D \ll \epsilon_F$ is always satisfied; consequently, we use the approximation $\int N(\xi) d\xi \simeq N(0) \int d\xi$, where

$$N(0) = \int \frac{d^3\mathbf{k}}{(2\pi)^3} \delta(\epsilon_F - \epsilon_k) = \frac{m k_F}{2\pi \hbar^2} \,, \tag{4.48}$$

is the density of states at the Fermi surface with $k_F = (3\pi^2 n)^{1/3}$ the Fermi wave number. In this way, the gap equation, Eq. (4.47), becomes

$$\frac{1}{|g|N(0)} = \int_0^{\hbar\omega_D} \frac{\tanh\left(\frac{\beta}{2}\sqrt{\xi^2 + \Delta^2}\right)}{\sqrt{\xi^2 + \Delta^2}}\, d\xi \, . \tag{4.49}$$

At zero temperature, from Eq. (4.49), we find immediately a simple formula for the zero temperature energy gap $\Delta(0)$:

$$\frac{1}{|g|N(0)} = \ln\left(\frac{\hbar\omega_D}{\Delta(0)} + \sqrt{1 + \frac{\hbar^2\omega_D^2}{\Delta(0)^2}}\right) . \tag{4.50}$$

Under the condition $\Delta(0) \ll \hbar\omega_D$, from Eqs. (4.50), we find the weak-coupling BCS result

$$\Delta(0) = 2\,\hbar\omega_D\, \exp\left(-\frac{1}{|g|N(0)}\right) \tag{4.51}$$

for the zero-temperature energy gap. The energy gap Δ decreases by increasing the temperature T and it becomes zero at the critical temperature T_c of the superconducting transition. Setting $\Delta(T_c) = 0$ in Eq. (4.49), we obtain

$$k_B T_c = 1.13\,\hbar\omega_D\,\exp\left(-\frac{1}{|g|N(0)}\right) . \tag{4.52}$$

Therefore, by considering the ratio between Eqs. (4.51) and (4.52), the zero-temperature energy gap $\Delta(0)$ in the weak-coupling regime is related to the critical temperature T_c as follows:

$$\Delta(0) = 1.764\, k_B T_c \, . \tag{4.53}$$

The results obtained here are valid in the deep BCS regime of metallic superconductors, where one can set $\Lambda = (2m\omega_F/\hbar)^{1/2}$ and $\mu = \epsilon_F$. However, in the case of ultracold fermionic atoms, the ultraviolet cutoff Λ cannot be easily identified with the inverse of a characteristic minimal length. In addition, in the BCS-BEC crossover, the chemical potential μ depends not only on the number density but also on temperature and interaction strength.

BCS-BEC Crossover for Ultracold Atoms

The crucial issue for the study of the BCS-BEC crossover with ultracold atoms is the observation that inserting Eq. (4.34) into Eq. (4.47) one finds

$$\frac{m}{4\pi\hbar^2 a} = \frac{1}{L^3}\sum_{|\mathbf{k}|<\Lambda}\left[\frac{1}{E_k}\tanh\left(\frac{\beta E_k}{2}\right) - \frac{m}{\hbar^2 k^2}\right], \tag{4.54}$$

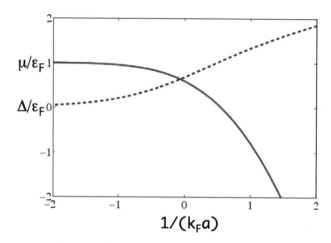

Fig. 4.3 Chemical potential μ and energy gap Δ in the BCS-BEC crossover at zero temperature as a function of the adimensional interaction strength $y = 1/(k_F a)$. The energies are in units of the Fermi energy $\epsilon_F = \hbar^2 k_F^2/(2m)$ with k_F the Fermi wave number

that is a renormalized gap equation valid both for $a < 0$ and $a > 0$. Remarkably, this equation remains finite and meaningful also in the limit $\Lambda \to +\infty$ because the two diverging terms in Eq. (4.54) compensate each other. As first discussed in [17], analyzing the renormalized gap equation, Eq. (4.54), coupled to the number equation, Eq. (4.46), one can investigate theoretically the BCS-BEC crossover. From this point of view, Eqs. (4.46) and (4.54) represent an extension of the BCS theory to include the full BCS-BEC crossover.

In Fig. 4.3 zero-temperature results for the chemical potential μ and the energy gap Δ obtained by solving numerically the coupled number equation, Eq. (4.46), and the renormalized gap equation, Eq. (4.54), both in units of the Fermi energy ϵ_F of the corresponding noninteracting gas, are reported versus the dimensionless parameter $y = 1/(k_F a)$, with $k_F = (3\pi^2 n)^{1/3}$ the Fermi wave number of noninteracting fermions.

Three distinct regimes can be identified, depending on the values of the parameter y:

(a) If $y \ll -1$, $\Delta/\epsilon_F \to 0$, while $\mu/\epsilon_F \to 1$. This is the deep BCS regime, in which there is a weak attraction, a size of the Cooper pair much larger than the average distance between any two electrons $\simeq k_F^{-1}$, and small deviations from the chemical potential of a noninteracting Fermi gas. Due to the small energy gap, we expect the BCS state to be quite fragile to temperature effects. More interestingly, an analytical approximation to the energy gap is

$$\Delta \simeq \frac{8}{e^2} \epsilon_F \exp[-\pi/(2k_F a)], \qquad (4.55)$$

also showing its nonanalytical dependence upon the scattering length.
(b) If $y = 0$, i.e., the scattering length diverges, we have a strongly interacting gas. The chemical potential is still positive, so the gas still behaves as a Fermi gas, and its value is close to the value of the energy gap. Quantitatively, one finds $\mu \simeq 0.59\,\epsilon_F$ and $\Delta \simeq 0.69\,\epsilon_F$.
(c) If $y \gg +1$, the chemical potential becomes negative and large in absolute value, $\mu \to -\epsilon_B/2$ with ϵ_B the binding energy of the molecular bosonic dimers made of atomic fermions with opposite spin. As we will see in this regime, the condensate fraction approaches the ideal Bose gas value, i.e., 100%, with all pairs (molecular dimers) in the condensate. This is the situation more robust with respect to finite temperature, and we expect critical temperatures for superfluidity of the order of the critical temperature for Bose-Einstein condensation of the molecular dimers.

This discussion also clarifies the difference between a BCS state and a BEC state, in that the BCS state needs first of all the formation of correlated pairs of fermions, while fermions are already correlated in a strong bound state in the BEC case. In light of this, the proximity of the critical temperature for superfluid ^{4}He and for the first superconductors discovered is absolutely accidental. In this second case, the outcome is due to the presence of a large Fermi temperature (order of 10^3–10^4 K for all metals) multiplied by a suppression factor, nonperturbative in the interaction strength, as the last term appearing in Eqs. (4.52) and (4.55) for superconductors and ultracold fermions, respectively. In the case of superconductors, this results in critical temperatures of the order of 10–10^2 K. In the ultracold atom case, the exponential factor may be comparable to unity, at least in the situations explored so far, and in this sense, superfluids made of fermions with BCS coupling mimic the behavior expected for "very high-temperature" superconductors.

In the BCS approximation for a uniform gas, we have introduced the energy gap Δ. This is a particular case of the more general definition of a pairing field $\Delta(\mathbf{r})$ as in Eqs. (4.19), (4.20), and (4.22). A nonzero value of this quantity implies the formation of Cooper pairs (two fermions with opposite spins) which give rise to a quasi-bosonic particle of zero spin zero. It seems natural, as first proposed by Chen Ning Yang [33], to consider this quantity a particular case of a spatial wave function for the two correlated fermions in the Cooper pair or as a density matrix for the Cooper pair as a whole

$$\Xi(\mathbf{r}, \mathbf{r}') = \langle \hat{\psi}_\downarrow(\mathbf{r}) \hat{\psi}_\uparrow(\mathbf{r}') \rangle, \quad (4.56)$$

such that $\Delta(\mathbf{r}) = -g\,\Xi(\mathbf{r}, \mathbf{r})$, i.e., is the diagonal elements of a density matrix. Evidently, if such a quantity can be evaluated, then the total number of Cooper pairs N_0 will be obtained as

$$N_0 = \int_{L^3 \times L^3} |\Xi(\mathbf{r}, \mathbf{r}')|^2 \, d^3\mathbf{r} \, d^3\mathbf{r}'. \quad (4.57)$$

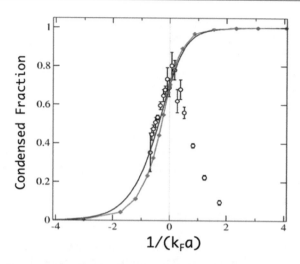

Fig. 4.4 Solid line: condensate fraction of Cooper pairs as a function of the adimensional interaction strength $y = 1/(k_F a)$, with k_F the Fermi wave number and a the s-wave scattering length. Circles with error bars: data of the same quantity as obtained in [36]. Solid line with diamonds: the condensate fraction computed in the local density approximation, $\mu \to \mu - U(\mathbf{r})$, and using Eqs. (4.46), (4.54), and (4.58), plotted against the value of y at the center of the trap. Credit: Reprinted figure with permission from [23]. Copyright 2005 by the American Physical Society

A finite value for the condensate fraction $N_0/(N/2)$ in the thermodynamic limit $N \to \infty$ implies the macroscopic occupation of a two-particle quantum state. This is the fermionic analog of the bosonic macroscopic occupation of a single-particle quantum state. Taking into account that in the uniform case $\Xi = -\Delta/g$, with the help of the gap equation (4.45), one finds that the condensate number of Cooper pairs satisfies this expression:

$$N_0 = \sum_\mathbf{k} \frac{\Delta^2}{4E_k^2} \tanh^2\left(\frac{\beta E_k}{2}\right). \qquad (4.58)$$

Studying Eq. (4.58) at zero temperature, one finds that the condensate fraction is exponentially small in the deep BCS regime, where a_F is very small and negative [23]. However, at unitarity the condensate fraction becomes quite large, and eventually, one gets $N_0/(N/2) = 1$ in the deep BEC limit for $a_F \to 0^+$.

This is shown in Fig. 4.4, together with experimental data taken with 6×10^6 ^{6}Li atoms confined in a anisotropic trap with cylindrical symmetry and large aspect ratio, $\omega_\perp/\omega_z = 47$, at the deepest Fermi degeneracy achievable ($k_B T/\epsilon_F = 0.05$) [36]. The agreement with the experimental data on the BCS side of the unitarity limit is good, while on the BEC side of the crossover, there is a rapid drop in the experimentally determined condensate fraction due to inelastic losses.

4.2 BCS Model

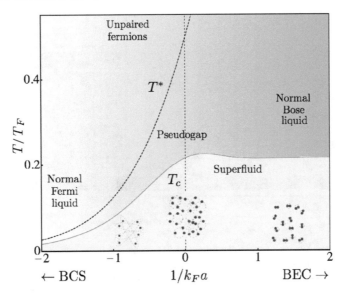

Fig. 4.5 Phase diagram of the BCS-BEC crossover in the $y - T/T_F$ plane. The dashed line gives the BCS critical temperature T^* of the superfluid-normal phase transition. The solid line is instead the critical temperature T_c obtained with a more sophisticated many-body approach. An intuitive sketch of the fermion pairing in position space for the three regimes appears at the bottom. Credit: reprinted figure with permission from [21], courtesy of the authors

4.2.2 Phase Diagram and Quantized Vortices

An accurate theoretical description of finite-temperature effects in the BCS-BEC crossover requires sophisticated many-body techniques with respect to the BCS approximation of Eq. (4.24), in particular in the BEC region close to the crossover. The predicted phase diagram in the plane $(y, T/T_F)$ is shown in Fig. 4.5. The dashed line gives the BCS critical temperature T^* of the superfluid-normal phase transition, while the solid curve gives a more accurate determination of the critical temperature T_c based on the Nozieres-Schmitt-Rink approach [19]. Notice that both T_c and T^* are obtained setting $\Delta = 0$ in the two theories. The pseudo-gap region is the one in between T_c and T^* with a positive chemical potential μ, which ensures the existence of a Fermi surface. In the pseudo-gap region, the energy gap Δ obtained with the extended BCS model is different from zero while is zero according to the Nozieres-Schmitt-Rink model. It is important to observe that also at zero temperature the predictions of the Nozieres-Schmitt-Rink formulation do not coincide with the ones of the extended BCS theory. For instance, in the BEC regime, the s-wave scattering length a_B of the bosonic molecules is given by $a_B = 2a_F$ according to the BCS theory while it is $a_B = 0.60 a_F$ within the Nozieres-Schmitt-Rink scheme. The experimental data have confirmed that the extended BCS approach is not fully reliable in the BEC regime.

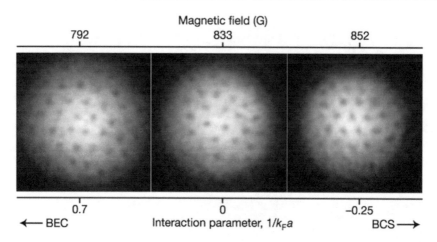

Fig. 4.6 Vortex lattices of a rotating cloud of ^{6}Li atoms for three values of the adimensional interaction strength $y = 1/(k_F a)$, changed by means of an external magnetic field close to the Feshbach resonance, with the value of the magnetic field indicated above each panel. The central panel is relative to the formation of vortices in the unitarity regime. Credit: reprinted figure with permission from [37]. Copyright 2004 by Springer Nature

As previously mentioned, the full BCS-BEC crossover was experimentally observed in 2003 with ultracold and dilute alkali-metal atoms. This groundbreaking experimental achievement was obtained by using the technique of Feshbach resonances, see Appendix C, where the scattering length a can be tuned at will by an external constant magnetic field **B**, including large positive and negative values close to resonance. While some evidences for superfluid states strongly rely on quantitative assessments and can sometime be mimicked by a classical gas, a qualitative smoking gun for superfluidity is the formation of vortices and vortex lattices, as in the case of Bose-Einstein condensates.

Quantized vortices have indeed been observed with ^{6}Li atoms for a broad range of the parameter y encompassing the three regimes, BCS, crossover, and BEC, described above. In particular, the formation of vortex lattices near the unitarity limit below a critical temperature was reported [37]. The measured critical temperature T_c at unitarity was around $0.2\epsilon_F/k_B$, in agreement with the Nozieres-Schmitt-Rink prediction. In Fig. 4.6, we show time-of-flight images of the ^{6}Li atoms after imprinting a mechanical rotation with blue-detuned focused laser beams. The three pictures correspond to different values of the s-wave scattering length a, controlled by an external uniform magnetic field. Systematic studies for the efficiency of vortex formation and their decay rate have also been performed, one example being reported in Fig. 4.7.

The study of superfluidity of fermions has required the refinement of techniques used for their bosonic counterpart or development of novel techniques. Among these, we mention three significant examples below.

4.2 BCS Model

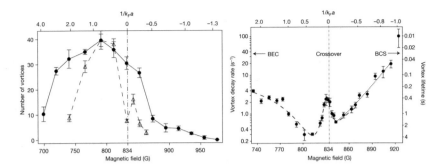

Fig. 4.7 Vortex formation and decay in a rotating cloud of ^{6}Li atoms. The left plot shows the number of vortices created versus the magnetic field (bottom horizontal axis) and therefore the corresponding y parameter (top horizontal axis). The maximum efficiency is achieved in the BEC regime close to unitarity. The right plot shows the dependence upon the same quantities of the vortex decay rate, which presents a local maximum at unitarity. Credit: reprinted figure with permission from [37]. Copyright 2004 by Springer Nature

Absorption and dispersive imaging techniques, already important to study collective excitations and the dynamics of phase separations for Bose gases, here require a careful handling due to the simultaneous presence of high magnetic fields if the gas is studied near the typical Feshbach resonances. Unlike Bose gases, with the exception of the study of spinor condensates, with trapped fermions, we are always dealing with at least two different hyperfine states when studying Cooper pairing. The different index of refraction for the two hyperfine states allows for their discrimination by using a double-shot image, spaced by a short time lapse, order of 10 μs, to avoid blurring due to photon recoil of the cloud imaged as second one, especially crucial for absorption imaging. In alternative, phase-contrast imaging can be used to provide directly the difference between the column densities of the two hyperfine states by proper choice of the detuning of the laser beam, without the need for subtraction of the separate images. These procedures are especially important to investigate unbalanced Fermi mixtures, in which Cooper pairing occurs partially until a regime is entered in which this is completely suppressed.

A second technique especially perfected for Fermi gases is the manipulation and measure of the atomic clouds with radiofrequency signals. These are important first of all to create a mixture of two hyperfine states. With respect to optical pumping at the level of the MOT trapping, a radiofrequency antenna offers advantages such as the negligible momentum transfer to the atoms and the equal addressability of the cloud due to a practically uniform RF magnetic field along the size of the cloud. A $\pi/2$ Rabi pulse will then create a coherent superposition with equal weights of two hyperfine states for a radiofrequency chosen as resonant with the transition between the two hyperfine states. The coherence is detrimental for the goal of having s-wave collisions leading to Cooper pairing, and a mechanism to decohere the cloud is therefore necessary. This is achieved by using the strong magnetic field curvature of the trapping potential, which imprints inhomogeneous phases to the atoms orbiting into different trajectories. Moreover, it is possible to perform radiofrequency spectroscopy to excite single atoms to infer informations about the binding energy in their BCS or BEC state. On the BEC side, this will correspond to measure the binding energy of the dimer, while on the BCS side, this will allow to measure the energy gap [27], in analogy to the tunneling experiments with superconductor-insulator junctions in solid-state physics.

The third technique concerns the momentum distributions in the BCS regime. Due to the fragility of the BCS state, this is achieved by converting the Cooper pairs into tightly bound molecules with a fast ramp of the magnetic field from the BCS side of the Feshbach resonance to a zero value [22, 36]. This is a subtle procedure, and the measured distribution of momenta on the BEC, molecular side is faithful to the original momentum distribution of the Cooper pairs if the ramp timescale is short enough before momentum changes occurs in the resulting molecular state, but long enough to allow the conversion of the Cooper pairs into molecules. The abundance of involved timescales in this process required a careful scrutiny and a close cooperation between theory and experiment.

The success of ultracold fermions in experimentally studying the BEC-BCS crossover should be soon followed by the study of more complex bosonization patterns, for instance, superfluidity with pairing in other than s-wave states and the so-called "exotic" superfluidity states. Two relevant examples are, respectively, states with nonzero momentum of the Cooper pair, the LOFF states [7, 15], and pairing in which one component remains gapless on the Fermi surface [18]. Also, while spinless bosons are only characterized by a simple $U(1)$ symmetry for the order parameter, fermions lend themselves to have more complex symmetries including their spin and angular momentum content. This is mapping the rich phase diagram of ^{3}He, but in a more controllable setting and with possibility of extensions to cases of interest in quantum field theory. Evidencing these superfluid states could also improve our understanding of the interplay between superconductivity and magnetism, a topic also of broad technological relevance [24].

4.3 Unconventional Phase Transitions

The conventional phase transitions usually follow the Landau paradigm in which an order parameter is introduced, manifesting a quantitative change, for instance, from a disordered phase at higher temperature in which it has zero value to a more ordered phase at lower temperature in which instead it assumes a value different from zero. This approach, inspired by phase transitions and critical phenomena in condensed matter systems and even applicable to Bose-Einstein condensation of an ideal gas, has been shown to be too restrictive in many contexts and has been subject to scrutiny in all its aspects. The outcome of this analysis shows that there are phenomena which can be considered as phase transitions even if they do not fulfil the Landau paradigm. In particular, there are phase transitions which are not driven by controlling the temperature or do not have an order parameter in the first place or only occur in dynamical, time-dependent situations. While the study of these phase transitions is still in progress, we discuss in the following four prototypical examples of unconventional phase transitions in three, two, and one spatial dimensions and, in

4.3.1 Mott-Superfluid Quantum Phase Transition

We discuss here a model in which the temperature does not play a role in determining the phase transition, rather another parameter is responsible for different phases. The simplest situation is perhaps the one in which one considers a system at zero temperature, such that thermal fluctuations are frozen out and cannot play a role in determining the evolution of the system. Then, the only way to induce macroscopic changes is to rely on quantum fluctuations triggering the change between two possible ground states of a many-body system. In the presence of a structural change of the many-body system, we call this phenomenon a quantum phase transition. The two ground states are related to the presence of two competing terms in the energy of the system and correspond to the dominance of one term over the other while changing a "control" parameter playing the same role of temperature for conventional, Landau-like, phase transitions. We recall that also in the case of temperature-driven phase transitions usually there is competition between two terms, in general the internal energy of a system and its entropy. The free energy $F = E - TS$ can be minimized either by having configurations with the lowest internal energy and low entropy or having configurations with excited states and, due to the increased multiplicity of the macrostate, with high entropy. In this setting, the temperature decides which one is the true ground state of the system.

In a more quantitative setting, let us consider the following second quantized Hamiltonian operator for a lattice containing bosons:

$$\hat{H} = -J \sum_{<i,j>} \hat{a}_i^+ \hat{a}_j + \sum_i \epsilon_i \hat{n}_i + \frac{1}{2} U \sum_i \hat{n}_i (\hat{n}_i - 1) \qquad (4.59)$$

also named Bose-Hubbard Hamiltonian, where $\hat{a}_i$ and $\hat{a}_i^+$ are the annihilation and creation operators for Bose particles in the lattice site i, such that the corresponding number operator is $\hat{n}_i = \hat{a}_i^+ \hat{a}_i$ (see also Appendix A). The particles experience an on-site repulsive interaction (if $U > 0$) and can hop from site i to a next-neighbor site j with a probability proportional to J. In the presence of an external potential, a term proportional to the single-particle potential energy in site i, and their number operator $\hat{n}_i$, will be also present. We notice that we have all the ingredients for the description of a quantum phase transition in an atomic physics setting. Let us suppose initially that no external potential is present ($\epsilon_i = 0$). The two competing terms in the dynamics are the hopping of bosons between sites (directly proportional to J, for instance, determined by quantum tunneling) and the repulsive interaction between on-site bosons (directly proportional to U, for instance, determined by a positive elastic scattering length). If the first term dominates ($U/J \to 0$), then we expect a ground-state single-state spread out over the entire lattice. In the opposite situation, we expect that the ground state is instead achieved with the atoms all localized in equal number n over all the sites. The two many-body ground states will be described, respectively, as

$$|\Psi\rangle_J \propto \left(\sum_{i=1}^{M} \hat{a}_i^\dagger \right)^N |0\rangle \tag{4.60}$$

$$|\Psi\rangle_U \propto \Pi_{i=1}^{M} (\hat{a}_i^\dagger)^n |0\rangle \tag{4.61}$$

representing, in the framework of second quantization, the two distinct representations in terms of Bloch and Wannier states. The presence of an external potential does not modify drastically this description but will be useful to deal with the experiment described in the following considerations (Fig. 4.8).

A clear demonstration of the dynamics of the Hamiltonian in Eq. (4.59) has been obtained using Bose-Einstein condensates of ^{87}Rb atoms created in a magnetic trap and then adiabatically transferred into a three-dimensional optical lattice [8]. The number of atoms in the condensate, with no discernible thermal component, was about 2×10^5, low enough to target an average number of atoms in each site of the optical lattice of order of few units. The optical lattice was formed by focusing three orthogonal laser beams, all at a wavelength of 852 nm, on the center of the magnetic trap. The condensate, of approximate spherical shape before the loading into the optical lattice, fragments itself into about 65 lattice sites in each dimension, therefore about 1.5×10^5 lattice sites. Due to the presence of the spatial modulation of the Gaussian beams and the simultaneous presence of a

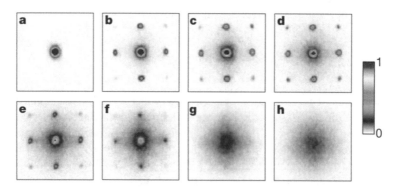

Fig. 4.8 Interference patterns after release of atoms from a three-dimensional optical lattice. The different absorption images, taken after a time of flight of 15 ms, correspond to different potential depths V_0 of the optical lattice, starting from $V_0 = 0$ (top left, no lattice, only harmonic potential of the magnetic trap) to the maximum value of $V_0 = 20E_r$ where E_r is the recoil energy (lower right), with the intermediate cases having $V_0 = 3, 7, 10, 13,$ and 16 times the recoil energy, from top left to lower right. The phase coherence of the atoms gets degraded by increasing the potential depth as visible by the reduced visibility of the blobs of constructive interference. When the potential depth is quite large, as in the last two images, coherence is completely lost, setting the transition at a value of V_0 intermediate between 13 and 16 always in units of recoil energy. Credit: reprinted figure with permission from [8]. Copyright 2002 by Springer Nature

weaker trapping potential, this ends up with about 2.5 atoms per lattice site in the center of the trap as peak average number. The phase coherence of the atoms in the whole lattice is tested by switching off the lattice, and look for the presence of three-dimensional interference patterns with wavelength determined by the lattice periodicity. The transition between the two phases is theoretically expected to occur at the ratio $U/J = 5.8z$, where z is the coordination number of the lattice, the number of next neighbohrs, in this case of a cubic lattice ($z = 6$) this ratio is about 36. Band structure calculations of U and J in terms of the potential depth and the lattice size yield a value of $U/J \sim 36$ when the potential depth is $V_0 \sim 13E_r$, consistent with the observations.

The reversibility of the interference fringes has been demonstrated by ramping up and down the potential depth. Finally, the jump from gapless excitations in the superfluid region to a finite gap in the Mott insulator region has been determined by using a potential with a gradient. The value of the gap is in good agreeent with the expectations of the Bose-Hubbard model. In the simplest case of a lattice with unity filling, the lowest energy excitation consists in transferring one atom on an adjacent site, leaving a hole. Due to the on site interaction, the price to be paid for this change is the repulsive energy U. This will inhibit tunneling between sites, unless energy is provided with a spatially dependent potential in such a way to compensate for the gap U.

The Bose-Hubbard model has been first introduced in the framework of liquid helium, and experimental studies have been carried out on solid-state systems such as granular superconductors and Josephson junction arrays. The advantages of the use of ultracold atoms are particularly evident here, due to the absence of several concomitant effect, impurities, and the possibility to use interferometry as an alternative to transport studies. In some sense, this experiment also opened the path to consider ultracold atoms, especially in optical lattices, as analog computers to simulate model Hamiltonians that are hard to solve analitically or with digital computers, as we will discuss in more detail in Chap. 5.

4.3.2 Berezinskii-Kosterlitz-Thouless Phase Transition

As described earlier, the interplay between Bose-Einstein condensation and superfluidity is not trivial, with an ideal Bose gas not being superfluid, for instance, based on the rather optimistic bound to the critical velocity given by the Landau criterion. The reverse can also occur if one considers Bose systems with lower effective dimensionality. As discussed earlier, BEC is not expected for a homogeneous Bose gas in two dimensions. This result is common with several phase transitions inhibited, in lower dimensionality, by the more important role played by the thermal fluctuations, responsible for disorder, with respect to the interaction between the microscopic agents leading to more ordered states. Nevertheless, in 1972, Kosterlitz and Thouless suggested that a 2D fluid can become superfluid below a critical temperature, the so-called Berezinskii-Kosterlitz-Thouless critical temperature T_{BKT}.

The original analysis started from the 2D XY model, which was originally used to describe the magnetization in a planar lattice of classical spins. The energy of the continuous 2D XY model is given by

$$E = \int \frac{J}{2} (\nabla \theta)^2 \, d^2\mathbf{r},$$

where $\theta(\mathbf{r})$ is the orientation of the magnetization field in each site and J is the *phase stiffness*, also named rigidity, related to the energy required to twist the spins by a given angle. This energy has the same expression of a complex scalar field:

$$\psi(\mathbf{r}) = \psi_0 \, e^{i\theta(\mathbf{r})} \tag{4.62}$$

of 2D superfluids with a uniform modulus $\psi_0 = \sqrt{n_{s0}}$, where $J = n_{s0}(\hbar^2/m)$ and neglecting the bulk energy.

A simple way to estimate the Berezinskii-Kosterlitz-Thouless (BKT) critical temperature T_{BKT} is to consider the Helmholtz free energy F of the 2D superfluid system characterized by one vortex line with topological charge $q = 1$ at temperature T. The internal energy E coincides with the kinetic energy of the vortex, which is obtained by integrating over the distance from the vortex eye between its core r_0 and the confinement radius of the system R:

$$E = \pi J \ln\left(\frac{R}{r_0}\right), \tag{4.63}$$

with $J = \hbar^2 n_{s0}/m$ the phase stiffness. The vortex can be formed around any of the points inside the circle of radius R, and considering that r_0 is the minimum distance to discern two vortex eyes, we get the entropy as of purely geometric character, from the number of possible configurations $(\pi R^2)/(\pi r_0^2)$, as

$$S = 2 k_B \ln\left(\frac{R}{r_0}\right). \tag{4.64}$$

It follows that the free energy can be written as

$$F = E - TS = (\pi J - 2 k_B T) \ln\left(\frac{R}{r_0}\right). \tag{4.65}$$

This function does not have singularities at finite r_0 and R. The temperature T_c at which F changes sign is

$$k_B T_c = \frac{\pi}{2} J = \frac{\pi \hbar^2 n_{s0}}{2m}. \tag{4.66}$$

For $T < T_c$, the free energy F is positive, and in the limit $R \to +\infty$, it goes to $F \to +\infty$. Instead, $T > T_c$ the free energy F is negative and in the limit $R \to +\infty$ it goes to $F \to -\infty$. Therefore, in the case of a very large system, T_c signals a

4.3 Unconventional Phase Transitions

topological phase transition from a superfluid phase ($0 \leq T < T_c$) without free vortices to a normal phase ($T > T_c$) characterized by the presence of free vortices.

This is the simplest argument for the Kosterlitz-Thouless transition based on a single vortex. Actually, Eq. (4.66) gives the same critical temperature T_{BKT} that one may derive from a more sophisticated approach based on the renormalization group.

A more refined analysis of Kosterlitz and Thouless based on the renormalization group shows that:

- As the temperature T increases, vortices start to appear as vortex-antivortex pairs (mainly with $q = \pm 1$).
- The pairs are bound at low temperature until at the critical temperature $T_c = T_{BKT}$ an unbinding transition occurs above which a proliferation of free vortices and antivortices is predicted.
- The phase stiffness J is renormalized by the presence of vortices.
- The renormalized superfluid density $n_s = J(m/\hbar^2)$ decreases by increasing the temperature T and jumps to zero above $T_c = T_{BKT}$.

An important prediction of the Kosterlitz-Thouless transition is that, contrary to the 3D case, in 2D the superfluid fraction n_s/n jumps to zero with a discontinuity above a critical temperature; see Fig. 4.9. For 3D bosonic superfluids, the transition

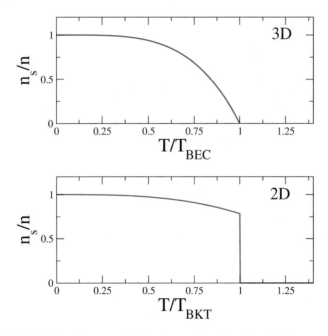

Fig. 4.9 Typical behavior of the superfluid fraction n_s/n as a function of the temperature T for a three-dimensional (3D) and two-dimensional (2D) superfluid

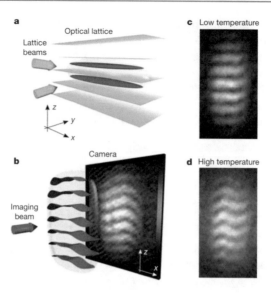

Fig. 4.10 Interference between two quasi-2D Bose gases at different temperatures, generated by turning on an optical lattice on a preexisting 3D Bose gas (**a**). The system is then illuminated with resonant light and the absorption image taken on the camera (**b**). There is a distinction between interference fringes taken at low temperature (**c**) and the ones taken at high temperature in which regular distorsions appear as expected by the presence of free vortices in the system. Credit: reprinted figure with permission from [9]. Copyright 2006 by Springer Nature

to the normal state is a BEC phase transition, while in 2D superfluids, the transition to the normal state is something different: a topological phase transition.

The initial field of applicability of the BKT model has been superfluidity in thin films of ^{4}He, superconductivity in 2D arrays of Josephson junctions, and collisional properties in 2D atomic hydrogen. However, the microscopic mechanism for the BKT phase transition, i.e., binding and dissociation of vortex-antivortex pairs, was not evidenced. This can instead be achieved by using quasi-2D quantum degenerate Bose gases [9]. The presence of different spatial correlation in the two phases has been investigated with an interferometric technique, therefore using two quasi-2D Bose systems, of which one is providing the reference for the interferometric signal. The schematics of the experiment is shown in Fig. 4.10. A 3D Bose gas is split into two planar systems by using an optical lattice potential with two laser beams intersecting at small angle. Thereafter, the trap is turned off and time-of-flight absorption images are taken, evidencing the existence of a relative phase between the two planar Bose gases. The study is repeated at different temperatures, and as seen in the images in (c) and (d), at low temperatures, the pattern is made of fringes as expected for coherent interference, while at high temperatures, there is a modulation of the pattern as expected by the presence in one of the two planar systems of vortices, inducing dislocations of the fringes.

4.3.3 Tonks-Girardeau Gas

We describe here a third example of strongly correlated systems still manifesting surprises, this time in one-dimensional Bose systems. The kinetic energy term per particle is written in general as $E_{kin} = \hbar^2 k^2/(2m)$. The wave vector $k = 2\pi/\lambda$ can be written in terms of the average interparticle distance, $\lambda \simeq \ell$ if the filling factor in an optical lattice is order of unity. In turn, the average interparticle distance can be written in terms of the atomic density as $\ell \sim n^{-1/D}$ in an optical lattice in D dimensions. Therefore, the average kinetic energy of an atom in the optical lattice will scale like $E_k \sim n^{2/D}$. This makes a qualitative distinction between 1 D and 3 D gases. The interaction energy is proportional in both cases to the density, while the kinetic energy scales as $n^{2/3}$ for a 3 D gas and as n^2 for a 1 D gas. Therefore, in the dilute limit, $n \to 0$, a 1 D gas will become more and more interacting, at variance with the 3 D case. Introducing a parameter γ as the ratio between interaction and kinetic energy per unit of particle, $\gamma = E_{int}/E_{kin}$, we have $I \to \infty$ as $n \to 0$.

For γ large, the effective repulsion between bosons becomes dominant and the ground state will be made of impenetrable boson. This is equivalent to the behavior of a corresponding ideal Fermi gas, mimicking their effective Pauli repulsion. It is therefore said that bosons in this regime are "fermionized." The ground-state many-body wave function at zero temperature is written as a Slater determinant: a strongly correlated 1D Bose gas is mapped into an equivalent noninteracting Fermi gas.

An experiment [20] using [87]Rb atoms has been performed the full Tonks-Girardeau regime of large γ by using an optical lattice with strong confinement in two directions; see example (a) in Fig. 2.9. This allows to reduce the dimensionality, as the dynamics along the directions of the strong confinement is frozen and results in strongly elongated condensates. In addition, an optical lattice was added along the elongated direction of the Bose gas. The atoms, subjected to the optical lattice potential, acquire a larger effective mass, and in doing so, the γ parameter becomes larger. With reference to the notation already used for the Mott-superfluid phase transition, if only the first Bloch band is occupied, the interaction and kinetic energy per unit of particle can be written as $E_{int} = Uv$, $E_{kin} = Jv$, respectively, with U the on-site interaction energy, J the tunneling amplitude, and v the filling factor, earlier assumeded to be of order unity. In this way, γ parameter between 5 and 200 has been achieved. The analogy to an ideal Fermi gas is however partial, as the momentum distribution is expected to differ. Even if in both cases the position distribution of the particles is similar, with all particles trying to be equally spaced from each other, fermions should also be spaced in momentum, while bosons can share the same momentum. The axial momentum distribution of the bosons can be measured by time-of-flight measurements of their spatial density after release from the optical and magnetic confinement, and found in agreement with the expectations for one-dimensional bosons until the axial lattice is so strong for the atoms to enter the Mott phase and fall back into the case of a tightly confined three-dimensional lattice. The request for low filling factors in this experiment implies the initial production of Bose condensate with small (3–4 $\times 10^4$) number of atoms already in the magnetic trap, affecting the precision in the determination of the momentum distribution.

4.3.4 Dynamical Critical Phenomena: Kibble-Zurek Mechanism

The fourth example of unconventional phase transition we consider deals with an extension of statistical mechanics to a system in which there is a sudden temporal variation of a parameter such that the system crosses a second-order phase transition, the so-called dynamical critical phenomena. The interest in these time-dependent phase transitions is rather broad and originated from the possible production of defects of various dimensionality in early cosmology [11, 12], with the proposal to used superfluid helium as an analog computer simulating this dynamics [34, 35]. In its simplest, classical version, one can consider a mean-field approach to a Gaussian model. The relaxation time, i.e., the time required to reach equilibrium, is found to diverge at the critical temperature, the so-called critical slowing down. Since the correlation length ξ is also diverging, it is natural to define the divergence of the relaxation time in terms of the latter, static quantity as $\tau_r \propto \xi^z$, with z a dynamical critical exponent. Introducing the critical exponent ν such that $\xi \propto |T - T_c|^{-\nu}$, we have $\tau_r \propto |T - T_c|^{-\nu z}$.

In the quantum setting, such slowing down of the dynamics toward equilibrium has more intriguing consequences. The analysis can be applied to a Bose gas for the transition between normal and condensed phases and to the Fermi gas for the transition between the normal and the superfluid phase. Suppose we start with the normal phase, in which atoms do not have a well-defined coherence among themselves. The gas is then ramped across the phase transition, for instance, with a linear protocol such that $T(t) = T_c(1 - t/\tau_Q)$ for $-\tau_Q/2 < t < +\tau_Q/2$, with T the temperature of the gas, and τ_Q is a characteristic time for the ramp speed, called quench time. In this way, the temperature is equal to $1.5\, T_c$ before the quench starts at $t = -\tau_Q/2$, decreases linearly afterward, until it reaches $0.5\, T_c$ at time $+\tau_Q/2$. This picture is valid if the gas is capable to adiabatically follow the quench, in the sense that it maintains its parts at thermal equilibrium. However, since the relaxation time increases as one approaches the phase transition (at time $t = 0$), there will be a time interval for which such an equilibrium is not observed, that we are going to estimate. By considering the dimensionless parameter $\epsilon(t) = t/\tau_Q$, the quench rate will be given by $|\dot\epsilon|/|\epsilon| = 1/|t|$, which means that the instantaneous timescale at which the quench occurs is t. The relaxation time diverges as $\epsilon^{-z\nu}$; therefore, there is a stage during the quench in which the system is out of equilibrium. This will happen in a time interval $\hat{t} \sim (t/\tau_Q)^{-z\nu}$, implying $\hat{t} \sim \tau_Q^{z\nu/(1+z\nu)}$. Therefore, we can distinguish between three stages, an initial adiabatic stage, an impulsive one in between $(-\hat{t}, +\hat{t})$, and a final adiabatic stage. In the impulsive regime, the condensate does not have enough time to extend a unique phase across its overall size, and domains with different phases will develop. This means that topological defects such as solitons, in one dimension, or vortices, in two or three dimensions, can emerge.

Experiments are therefore aimed at measuring the scaling law of defects produced with the quenching rate. The mechanism has been already studied in different systems, ranging from superfluid ^{3}He to superconducting films including also rings and ion chains. In the case of Bose condensates, experiments have been performed to generate vortices [32], also in a toroidal trap [4], and solitons [14]. The crucial observable in these experiments is the scaling of the average number of solitons with the quench time. Successive experiments have been performed with the formation of vortices with degenerate Fermi systems in a single optical dipole trap [13] and in a toroidal trap [1].

4.4 Problems

4.1. Extend the Thomas-Fermi density functional to the time-dependent case adopting an hydrodynamic approach, where the dynamical fields are the local number density $n(\mathbf{r}, t)$ and the local velocity field $\mathbf{v}(\mathbf{r}, t)$.

4.2. Calculate the critical strength which gives rise to the Stoner instability for a two-spin component Fermi gas in D dimensions, with $D = 1, 2$.

4.3. Within the BCS formalism at zero temperature, find analytically the chemical potential, the energy gap, and the condensate fraction of the two-dimensional attractive Fermi gas in the BCS-BEC crossover.

4.4. In the framework of the BCS theory, determine the heat capacity of the attractive Fermi gas at the critical temperature.

4.5. Prove the decoupling approximation $\hat{A}\hat{B} \simeq \langle\hat{A}\rangle\hat{B} + \hat{A}\langle\hat{B}\rangle - \langle\hat{A}\rangle\langle\hat{B}\rangle$ for two generic operators $\hat{A}$ and $\hat{B}$ under the assumption that their fluctuations around the mean values $\langle\hat{A}\rangle$ and $\langle\hat{B}\rangle$ are small.

4.6. Considering the Bose-Hubbard model at zero temperature with U the onsite interaction strength and J the hopping energy, derive the regions of the diagram (U, J) where the system is superfluid.

4.5 Further Reading

- C. N. Yang, *Concept of Off-Diagonal Long-Range Order and the Quantum Phases of Liquid He and of Superconductors*, Rev. Mod. Phys. **34**, 694 (1962).
- O. Madelung, *Introduction to solid-state theory* (Springer-Verlag, Berlin and Heidelberg, 1978).
- A. J. Leggett, *Diatomic molecules and Cooper pairs*, in *Modern Trends in the Theory of Condensed Matter*, A. Pekalski A. and J. Przystawa editors, (Springer-Verlag, Berlin, 1980), pp. 14-27.
- A.J. Leggett, *Quantum Liquids: Bose condensation and Cooper pairing in condensed matter systems* (Oxford University Press, 2006).

- C. Chin, R. Grimm, P. Julienne, and E. Tiesinga, *Feshbach resonances in ultracold gases*, Rev. Mod. Phys. **82**, 1225 (2010).
- W. Zwerger (editor), *The BCS-BEC Crossover and the Unitary Fermi Gas* (Springer, Berlin, 2012).
- M. Lewenstein, A. Sampera, and V. Ahufinger, *Ultracold atoms in optical lattices* (Oxford University Press, 2012).
- D. Vollhardt and P. Wolfle, *Superfluid phases of Helium 3* (Dover, New York, 2013).
- M. Randeria and E. Taylor, *Crossover from Bardeen-Cooper-Schrieffer to Bose-Einstein Condensation and the Unitary Fermi Gas*, Annual Rev. Cond. Matt. Phys. **5**, 209 (2014)
- L. Salasnich and F. Toigo, *Zero-point energy of ultracold atoms*, Phys. Rep. **640**, 1 (2016).

References

1. Allman, D.G., Sabharwal, P., Wright, K.C.: Quench-induced spontaneous currents in rings of ultracold fermionic atoms. Phys. Rev. A **109**, 053320 (2024)
2. Bardeen, J., Cooper, L.N., Schrieffer, J.R.: Theory of superconductivity. Phys. Rev. **108**, 1175 (1957)
3. Bogoliubov, N.N.: On a new method in the theory of superconductivity. Nuovo Cimento **7**, 794 (1958)
4. Corman, L., Chomaz, L., Bienaimé, T., Desbusuois, R., Weitenberg, C., Nascinbène, S., Dalibard, J., Beugnon, J.: Quench-induced supercurrents in an annular Bose gas. Phys. Rev. Lett. **113**, 135302 (2014)
5. Eagles, D.M.: Possible pairing without superconductivity at low Carrier concentrations in bulk and thin-film superconducting semiconductors. Phys. Rev. **186**, 456 (1969)
6. Fermi, E.: Bestimmung einiger Eigenschaften des Atoms und ihre Anwendung auf die Theorie des periodischen Systems der Elemente. Z. Phys. **48**, 73 (1928)
7. Fulde, P., Ferrel, R.A.: Superconductivity in a strong spin-exchange field. Phys. Rev. A **135**, 550 (1964)
8. Greiner, M., Mandel, O., Esslinger, T., Hänsch, T.W., Bloch, I.: Quantum phase transition from a superfluid to a Mott insulator in a gas of ultracold atoms. Nature **415**, 39 (2002)
9. Hadzibabic, Z., Kruger, P., Cheneau, M., Battelier, B., Dalibard, J.: Berezinskii–Kosterlitz–Thouless crossover in a trapped atomic gas. Nature **441**, 1118 (2006)
10. Illuminati, F., Albus, A.: High-temperature atomic superfluidity in lattice Bose-Fermi mixtures. Phys. Rev. Lett. **93**, 090406 (2004)
11. Kibble, T.W.B.: Topology of cosmic domains and strings. J. Phys. A **9**, 1387 (1976)
12. Kibble, T.: Some implications of a cosmological phase transition. Phys. Re. **67**, 183 (1980)
13. Ko, B., Park, J.W., Shin, Y.: Kibble-Zurek universality in a strongly interacting Fermi superfluid. Nat. Phys. **15**, 1227 (2019)
14. Lamporesi, G., Donadello, S., Serafini, S., Dalfovo, F., Ferrari, G.: Spontaneous creation of Kibble-Zurek solitons in a Bose-Einstein condensate. Nat. Phys. **9**, 656 (2013)
15. Larkin, A.I., Ovchinnikov, Y.N.: Nonuniform state of superconductors. Zh. Eksp. Teor. Fiz. **47**, 1136 (1964) [Sov. Phys. JETP **20**, 762 (1965)]
16. Leggett, A.J.: Diatomic Molecules and Cooper Pairs. In: Pekalski, A., Przystawa, R. (eds.) Modern Trends in the Theory of Condensed Matter. Spinger, Berlin (1980)
17. Leggett, A.J.: Cooper pairing in spin-polarized Fermi systems. J. Physique. **41**, C7-19 (1980)
18. Liu, W.V., Wilczek, F.: Interior gap superfluidity. Phys. Rev. Lett. **90**, 047002 (2003)

References

19. Nozieres, P., Schmitt-Rink, S.: Bose condensation in an attractive fermion gas: From weak to strong coupling superconductivity. J. Low Temp. Phys. **59**, 195 (1985)
20. Paredes, B., Widera, A., Murg, V., Mandel, O., Folling, S., Shlyapnikov, G.V., Cirac, I., Hänsch, T.W., Bloch, I.: Tonks–Girardeau gas of ultracold atoms in an optical lattice. Nature **429**, 277 (2004)
21. Randeria, M., Taylor, E.: Crossover from Bardeen-Cooper-Schrieffer to Bose-Einstein Condensation and the Unitary Fermi Gas. Annu. Rev. Condens. Matter Phys. **5**, 209 (2014)
22. Regal, C.A., Greiner, M., Jin, D.S.: Observation of resonance condensation of fermionic atom pairs. Phys. Rev. Lett. **92**, 040403 (2004)
23. Salasnich, L., Manini, N., Parola, A.: Condensate fraction of a Fermi gas in the BCS-BEC crossover. Phys. Rev. A **72**, 023621 (2005)
24. Sarma, G.: On the influence of a uniform exchange field acting on the spins of the conduction electrons in a superconductor. Phys. Chem. Solids **24**, 1029 (1963)
25. Schafroth, M.R.: Superconductivity of a charged boson gas. Phys. Rev. **96**, 1149 (1954)
26. Schafroth, M.R.: Theory of superconductivity. Phys. Rev. **96**, 1442 (1954)
27. Schirotzek, A., Shin, Y., Schunck, C.H., Ketterle, W.: Determination of the superfluid gap in atomic Fermi gases by quasiparticle spectroscopy. Phys. Rev. Lett. **101**, 140403 (2008)
28. Stoner, E.C.: Atomic moments in ferromagnetic metals and alloys with non-ferromagnetic elements. Philos. Mag. **15**, 1018 (1933)
29. Thomas, L.H.: The calculation of atomic fields. Proc. Cambridge Philos. Soc. **23**, 542 (1926)
30. Valatin, J.G.: Comments on the theory of superconductivity. Nuovo Cimento **7**, 843 (1958)
31. Valtolina, G., Scazza, F., Amico, A., Burchianti, A., Recati, A., Enss, T., Inguscio, M., Zaccanti, M., Roati, G.: Exploring the ferromagnetic behaviour of a repulsive Fermi gas through spin dynamics. Nat. Phys. **13**, 704 (2017)
32. Weiler, C.N., Neely, T.W., Scherer, D.R., Bradley, A.S., Davis, M.J., Anderson, B.P.: Spontaneous vortices in the formation of Bose–Einstein condensates. Nature **455**, 948 (2008)
33. Yang, C.N.: Concept of off-diagonal long-range order and the quantum phases of liquid He and of superconductors. Rev. Mod. Phys. **34**, 694 (1962)
34. Zurek, W.H.: Cosmological experiments in superfluid liquid helium?. Nature **317**, 505 (1985)
35. Zurek, W.: Cosmological experiments in condensed matter systems. Phys. Rep. **276**, 177 (1996)
36. Zwierlein, M.W., Stan, C.A., Schunck, C.H., Raupach, S.M.F., Kerman, A.J., Ketterle, W.: Condensation of pairs of fermionic atoms near a Feshbach resonance. Phys. Rev. Lett. **92**, 120403 (2004)
37. Zwierlein, M.W., Abo-Shaeer, J.R., Schirotzek, A., Schunck, C.H., Ketterle, W.: Vortices and superfluidity in a strongly interacting Fermi gas. Nature **435**, 1047 (2005)

Ultracold Atoms as Coherent Systems 5

One of the most important properties of ultracold atoms is their high degree of quantum coherence. The de Broglie wavelength is associated to the whole atomic sample, as already discussed especially in Chap. 3. Experimental evidences for this feature have been shown both in the direct observation of interference fringes from two condensates and in the manipulation of the phase of the condensate to imprint velocity fields, as demonstrated in the production of solitons and vortices. In this chapter, we focus on quantum coherence of ultracold atoms and its impact in a variety of applications to both measurements of fundamental physics and engineering, most notably in analog and digital computing. One of the common features of these applications is that interatomic interactions are generally a nuisance; therefore, it is convenient to minimize them as much as possible. This also completes the big picture, as in Chap. 3 we have instead discussed weak interatomic interactions and in Chap. 4 strong interatomic interactions. This chapter can then also be considered a return to noninteracting atoms as in the ideal setting presented in Chap. 1 but now enriched by an enabling technology of trapping, cooling, and manipulating atomic samples for metrological and computational goals.

More in detail, after some preliminary remarks on the concept of coherence, we start the discussion with one of the most important setups in experimental physics, the optical interferometer. The reader should have already been exposed to this tool at least in connection with the crucial experiment carried out by Michelson and Morley [37] and ruling out the ether as possible medium for the propagation of electromagnetic waves, paving the road to special relativity. We then proceed discussing atomic interferometers, always stressing the analogy to optical interferometers, and emphasizing advantages and disadvantages. Atomic interferometers already exist and do not use highly coherent atomic beams, like the original Michelson-Morley interferometer did not use highly coherent light beams. It is therefore natural to discuss the benefits of using high brightness, high coherence atomic beams as inputs, in analogy to the use of lasers for optical interferometers. This requires atom lasers, which have been demonstrated with various output coupling schemes. Lasing requires matter-wave amplification, and we then discuss experimental work in this direction as well. Detailed studies of matter-wave interference are also available with ultracold atoms in traps, and in particular, we describe the analog of Josephson effects. All these elements allow to discuss two active directions of research, the use of ultracold atoms as carriers in circuits to mimic electronic devices, the emerging subfield of engineering called atomtronics, and as analog platforms to simulate specific model Hamiltonians of quantum many-body physics, the so-called quantum emulators.

5.1 Coherence of Photons and Matter Waves

Before discussing atomic interferometry, it is important to clarify the meaning of the word "coherence," already used in an optical context introduced in a pure undulatory approach. In that setting, coherence is about the predictability of the phase of a wave at a later time or in another point different from the emission one. Perfect coherence means that such a phase is predictable for any point in space time. In practice, the mechanism to produce waves, the interactions encountered along the path, or even the very act of measurement can give rise to partial or complete lack of predictability. All the above have to be considered sources of "decoherence," i.e., loss of coherence. The simplest way to quantify the degree of coherence of a wave is to look at the intensity collected in a region and averaged over a time much longer than the oscillation period of the wave, after having traveled from two distinct paths. For a perfectly coherent situation, we will have regions in which there is complete constructive interference and regions in which there is complete destructive interference. If coherence is completely lost, we will observe a constant intensity along all the collection region, with no contrast whatsoever. In the most likely intermediate case, there will be some partial interference, with maxima and minima of intensity $I_{\max}$ and $I_{\min}$. We can define the fringe visibility as

$$V = \frac{I_{\max} - I_{\min}}{I_{\max} + I_{\min}}, \quad (5.1)$$

with $V = 0$ the case of complete decoherence, $V = 1$ complete coherence, and partial coherence/decoherence in all intermediate cases. More sophisticated indicators of coherence are the timescales and lengthscales on which predictability of the undulatory behavior is warranted. These are called coherence time and coherence length, respectively, showing that the same source of waves can be thought as coherent or decoherent depending on the timescale and lengthscale involved in the experimental setup. The realization that electromagnetic waves are made of energy "lumps," the photons, required a conceptual adjustment which is at the core of the probabilistic interpretation of quantum mechanics. The photon travels along various paths at the same time with various probabilities and is then detected as a whole particle. Dirac stressed this reinterpretation of interference with a famous sentence: "Each photon interferes only with itself. Interference between different

5.1 Coherence of Photons and Matter Waves

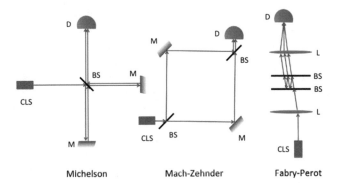

Fig. 5.1 Examples of optical interferometers. From left to right, Michelson, Mach-Zehnder, and Fabry-Perot interferometers. They differ by the path of the light beams, with only the ones relevant to the final interference evidenced. The various elements are a coherent light source (CLS), beam splitters (BS), mirrors (M), light intensity detector (D), and, in the case of the last, converging lenses (L)

photons never occurs" [15]. This strong statement is supported by performing experiments with photon flux so low that the simultaneous presence of two photons is highly improbable. In this situation, photon interference is confirmed to be a single particle affair. The advent of light sources with high degree of temporal and spatial coherence, together with high brightness—a measure of the intensity of the source in the direction of propagation of the wave—such as lasers, offered another perspective. In this case, photons produced by independent sources can interfere, as demonstrated in the microwave [33] and in the visible [32] regions of the electromagnetic spectrum. Ultimately, the interference occurs in the amplitude of general quantum mechanical states, and these can be made of single photons or multiphotons alike, so the problem is somewhat ill posed [20] (Fig. 5.1).

Unfortunately, the word coherence also appears to indicate another concept, the one of a "coherent" state, as defined in Appendix B. This is a particular class of quantum states first introduced by Schrödinger to describe the dynamics of a harmonic oscillator in analogy to its classical limit. Since electromagnetic waves can be represented as a system of infinite harmonic oscillators, this is also the more natural class of quantum states with immediate classical limit. Even if on average these states mimic classical mechanics, they continue to be quantum states with a peculiar signature in their fluctuations around the average. These states have minimum uncertainty in the phase ϕ and large uncertainty in the photon number N. They are therefore at variance with "number" states in which vice versa the number of photons is well defined at the price of large uncertainties on the phase, such that $\Delta N \Delta \phi \geq 1$ is always enforced [7]. Coherent states of light, for which the phase is defined, have therefore large photon number fluctuations, while number states have corresponding large phase fluctuations. As long as photons are considered, large number fluctuations for coherent states do not create potential issues, since we are accustomed to the idea that photons, with zero chemical potential, can be easily created and destroyed. How about repeating these considerations for matter waves, in which we definitely start with, at least in principle, well-defined number of particles? The problem became even more acute after the realization of the interference experiment with Bose condensates described in Chap. 3.

The solution has been found by describing the detailed dynamics of formation of coherence in an interference experiment between two Bose-Einstein condensates prepared in different states [8, 26]. The discussion in [8], to which we refer the reader for a complete account, considers a simplified version of an interference experiment, in which both condensates can tunnel with equal probability through a barrier and then are mixed into a 50-50 atomic beam splitter, and the atoms are thereafter separately detected with two ideal efficiency counters. This avoids the complications of discussing interference patterns in space but contains the same physics in terms of detection dynamics. If we start with a "coherent" state for each condensate, each of the two beams incident on the beam splitter will be described by macroscopic wave functions with defined phases, $|\psi_A\rangle = |\psi_0| \exp(-i\phi_A)$ and $|\psi_B\rangle = |\psi_0| \exp(-i\phi_B)$, where we suppose that the two beams incident on the beam splitter have the same average number of atoms and are characterized by two phases ϕ_A, ϕ_B which are uncorrelated and varying for each realization of the experiment. The beating of the two modes after the beam splitter will originate intensities detected at the two counters which will be $I_A = 2|\psi_0|^2 \cos^2 \phi$ and $I_B = 2|\psi_0|^2 \sin^2 \phi$, with $\phi = (\phi_A - \phi_B)/2$ spanning over a π range. As an example, the probability to detect a sequence of k atoms in counter A, and 0 in counter B in the same experiment, is simply $P(k, 0) = \cos^{2k} \phi$. The phase ϕ is unknown, and therefore, the average probability will be obtained by averaging $P(k, 0)$ over the entire range of possible ϕ

$$P^{\text{Coherent}}_{(k,0)} = \frac{1}{\pi} \int_{-\pi/2}^{\pi/2} d\phi \cos^{2k} \phi = \frac{(2k)!}{(2^k k!)^2} \sim \frac{1}{\sqrt{\pi k}} \quad (5.2)$$

where the last expression is obtained in the large k limit using the Stirling approximation for the factorials.

Now, we instead consider two condensates with well-defined number of atoms, say $N_A = N_B = N$, such that the initial quantum state is

$$|\psi_0(t_0)\rangle_{00} = |N_A\rangle \otimes |N_B\rangle = |N, N\rangle . \quad (5.3)$$

After tunneling and beam splitting, a first atom is detected in one of the two counters; therefore, the initial state of the condensates is decreased by one atom, which can be taken into account by using annihilation operators $\hat{a}$ and $\hat{b}$ acting on condensate A and B, respectively. In more physical terms, we do not know to which condensate the detected atom belonged, and this means that after its detection the quantum state of the two condensates is in one of two possible superposition states, orthogonal to each other

$$|\psi_0(t_1)\rangle_{10} = (\hat{a} + \hat{b})|\psi_0(t_0)\rangle_{00} = \sqrt{N}(|N - 1, N\rangle + |N, N - 1\rangle) , \quad (5.4)$$

5.1 Coherence of Photons and Matter Waves

if the atom is detected in counter A, and

$$|\psi_0(t_1)\rangle_{01} = (\hat{a} - \hat{b})|\psi_0(t_0)\rangle_{00} = \sqrt{N}(|N-1, N\rangle - |N, N-1\rangle), \quad (5.5)$$

if the atom is detected in counter B, with each detection having a probability 1/2 to occur.

These new states will affect the dynamics of the second atom to be detected. After the second detection, the quantum state of the two condensates will be one of the following states conditional to the detection of the first atom in counter A:

$$|\psi_0(t_2)\rangle_{20} = (\hat{a} + \hat{b})|\psi_0(t_1)\rangle_{10} = \sqrt{N(N-1)}\,(|N-2, N\rangle + |N, N-2\rangle)$$
$$+2N|N-1, N-1\rangle, \quad (5.6)$$
$$|\psi_0(t_2)\rangle_{11} = (\hat{a} - \hat{b})|\psi_0(t_1)\rangle_{10} = \sqrt{N(N-1)}(|N-2, N\rangle - |N, N-2\rangle),$$

if the second atom is also detected in counter A and if it is instead detected in counter B, respectively. If instead the first atom has been detected in counter B, the states after the second detection will be

$$|\psi_0(t_2)\rangle_{11} = (\hat{a} + \hat{b})|\psi_0(t_1)\rangle_{01} = \sqrt{N(N-1)}(|N-2, N\rangle - |N, N-2\rangle),$$
$$|\psi_0(t_2)\rangle_{02} = (\hat{a} - \hat{b})|\psi_0(t_1)\rangle_{01} = \sqrt{N(N-1)}\,(|N-2, N\rangle + |N, N-2\rangle) \quad (5.7)$$
$$-2N|N-1, N-1\rangle,$$

if the second atom is detected in counter A and if is instead also detected in counter B, respectively. Notice that while the probability to detect two atoms in the same counter is proportional to $2N(N-1) + 4N^2$, the probability to measure the second atom in one counter (conditional to have measured the previous atom in the other counter) is instead proportional to $N(N-1)$. In the limit of large N, this means the somewhat counterintuitive result that the probabilities for detecting two atoms in one counter are three times the probability for detecting one atom in one counter and the successive atom in the other counter. The procedure can be iterated, and the probability that there will be k detection in one counter, and zero in the other, is found to be

$$P_{(k,0)}^{\text{Fock}} = \frac{1}{2}\frac{3}{4}\frac{5}{6}\cdots\frac{(2k-1)}{2k} = \frac{1 \cdot 2 \cdot 3 \cdot 4 \cdot 5\ldots \cdot (2k-1) \cdot 2k}{(2 \cdot 4 \cdot 6 \ldots \cdot (2k))^2} = \frac{(2k)!}{(2^k k!)^2}, \quad (5.8)$$

which coincides with $P_{(k,0)}^{\text{Coherent}}$ in Eq. (5.2) that is from a completely different initial state. This shows how coherence builds up even starting from states, such as number states, among the farthest from the specific class of "coherent" states. Indeed, it would be much less confusing to distinguish between "phase" states and "number" states, but as in other cases, the traditional term persists. In passing, we find another intriguing feature of matter waves. Due to their low velocity, atoms emitted from coherent (i.e., with long coherence times and large coherence lengths) sources will affect the density of the source while being still inside, as we will discuss for a striking and relevant example in Sect. 5.4.

The role of coherence is not limited to the observation of interference patterns with some visibility but also determines the fluctuations in phase and number of the atoms or photons. In this case, we speak about higher-order coherence, which is determined by the statistical properties of the source. Recalling the analogy with light, even if the interference fringes produced by a thermal light source and a laser source are identical, they will differ in the photon counting affecting then the shot-to-shot fluctuations. It is intuitive to think that, at least from entropic considerations, a laser source will have more predictable fluctuations than a thermal source. The average quadratic number fluctuations can be written as

$$\Delta n^2 = \langle (n - \langle n \rangle)^2 \rangle = \langle n^2 \rangle - \langle n \rangle^2 = (g^{(2)}(0) - 1)\langle n \rangle^2, \tag{5.9}$$

where we have introduced the second-order coherence function $g^{(2)}(r)$ of the distance r between two particles, evaluated in the limit of $r \to 0$. While for thermally fluctuating particles we expect $g^{(2)}(0) = 0$, the absence of density fluctuations corresponds to the less entropic situation, with $g^{(2)}(0) = 1$. When atoms detected with a counter, the difference in second-order coherence will be manifested by the statistics of the time between two consecutive detections. In the case of thermal atoms, this will follow a Poissonian distribution, while for atoms sourced from a Bose-Einstein condensate, the statistics will be sub-Poissonian or bunched, with more predictable detection times.

In general, for the coherence at the nth order, we will have $g^{(n)}(0) = 1$ for a Bose-condensate and $g^{(n)}(0) = n!$ for a thermal state. In the atomic case, measurements of the second- and third-order coherences have been performed by using indicators of the square power and the cubic power of the density, respectively. Then, $g^{(2)}(r)$ can be inferred by the mean-field energy of the atomic cloud, while $g^{(3)}(r)$ can be derived, in the limit of losses dominated by three-body recombination, by the lifetime of the atoms in the trap.

5.2 Atomic Interferometers

Atomic interferometers are the analog of optical interferometers; therefore, it is worth to discuss first examples of the latter and then translate their elements into the atomic realm. Among the optical interferometers, a first classification can be based on the paths followed by light. In some, one is interested to distinguish between phase differences induced by an external agent acting in different directions of space. This is the case of a Michelson-Morley interferometer. Having recognized that the ether, if existing at all, uniformly permeates the whole space, the only possibility to detect it was to look for differences in the velocity of light propagating in different directions. Starlight aberration supported the hypothesis that the earth is in motion with respect to the frame of reference in which the ether is at rest, with the possibility of a modulation of the signal based on rotation and revolution

5.2 Atomic Interferometers

of the earth. In other setups, the external agent is a medium which instead can be contained, for instance, in a small region of space, such as a cell filled with gas at various pressures in which the aim is to measure the index of refraction of the gas. In this case, it may be convenient to have two separate paths for light, sending only one beam on the cell filled with gas and the other on a cell in vacuum. Mach-Zehnder interferometers are of this kind. Moreover, one can also look for interference from multiple paths, which may be advantageous to filter different spectral components of light or to detect displacements of the end mirrors, as in Fabry-Perot interferometers. All these three interferometers and their combination (the Michelson-Morley interferometer also used multiple paths to enhance sensitivity) have counterparts in atomic interferometry, as we will describe soon. Also, it should be kept in mind that further classifications are possible, for instance, based on the use of internal states, such as two different energy levels of the atoms, or external states, such as different momentum or positional states. In this regard, the crucial feature is that the external agent to be detected should have different interactions (and dephasing) for the different states. Interferometers also may differ in the devices used to create different states, for instance, with use of nanofabricated gratings or using laser beams. Moreover, another classification consists in distinguishing between spatial and temporal interferometry, in the latter case the trajectory of the atoms is the same, but the timing of the interaction with light beams is different for the two paths. Given the variety of cases, we will limit the attention to some representative elements which should be basic enought to unpack the entire panoply of experimental setups.

It is important to mention from the outset that atomic interferometers are the more recent ones among interferometers based on matter waves. The first use of matter waves was pioneered by Davisson and Germer evidencing electron diffraction [13], with the following demonstration of electron interferometers [36], and has resulted in the development of low-energy electron diffraction (LEED) techniques now used for analysis of surface structures of crystals. Then, interference of neutrons beams was demonstrated in a Mach-Zehnder configuration [43] and used to detect the effect of gravity on the phase of neutrons, the first experiment in which gravitation and quantum mechanics were simultaneously involved [11]. Atoms present distinctive advantages with respect to electrons and neutrons, among these the flexibility offered by using a variety of atoms prepared in controllable internal states, the larger light scattering cross sections, easiness of production not requiring costly electron guns or nuclear reactors, and portability now at the level of miniaturization on atom chips.

The first element of an atomic interferometer is an atom source. As in the optical case, an ideal source would be one with the largest brightness and the more defined momentum and energy of the atoms. However, optical interferometry was developed well before the advent of laser sources; therefore, we expect that also in the case of atoms "thermal" sources, the analog of light lamps, should be enough. In addition, we need the analogous of beam splitters, diffraction gratings, and mirrors, keeping in mind that de Broglie wavelengths for typical atomic beams are in the nm range. Human-made nanostructures allow to design these elements, generally

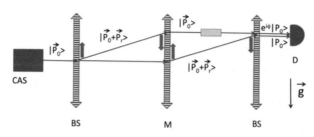

Fig. 5.2 Example of atomic interferometer. Atoms are emitted by a coherent atomic sources (CAS), and with a properly designed interaction with a laser beam, half of them acquire a momentum in the transversal direction, for instance, through Bragg scattering at first order, with two units of recoil momentum, $\mathbf{p}_r = 2\hbar\mathbf{k}_r$, emulating a beam splitter (BS). This separates the beam into two components which, after further recoils with the two laser beams used as mirrors (M) and as a recombiner (BS), are then measured by an intensity-sensitive detector. The possible presence of an external agent schematized here as a medium (in the top right path), or the very presence of the earth's gravitational field, will induce relative phase shifts in the two matter waves, affecting the intensity measured at the detector. In the presence of gravity, all trajectories are parabolic, replacing the straight lines depicted here

at the price of reduced intensities of the atomic beams, due to absorption effects, and dependence on the technological process of production, with the possibility of systematic effects due to defects and corrugations. We therefore limit the attention, also for reasons of space, to elements made with light beams, broadly used in the current generation of interferometers for metrological use (Fig. 5.2).

Absorption and spontaneous emission of light impart energy and momentum to atoms, as discussed in Chap. 2. A variant of this process consists in absorption and stimulated emission of light. This has the advantage that the resulting momentum change can be controlled in direction, but it requires a second light beam to "guide" the stimulated emission process in the desired direction. If this process does not change the internal state of the atom, this is called Bragg scattering; otherwise, it is called Raman scattering. This process is complementary to Bragg scattering of photons from matter already used in condensed matter physics to characterize spatial order. The role of atoms and light is simply reversed, except that in atomic interferometers the process is pulsed in time, and the matching angle is replaced by a condition on the detuning of the two laser beams.

Fundamental in this technique is also the duration of the laser pulse, which should be chosen, together with the laser intensity, to bring the atoms into a state which is a 50-50 superposition state or to bring the entire atomic population in the upper state (if the initial beam is in the lower state), exploiting the phenomenon of Rabi oscillations. In its simplest version, this consists in the dynamics involved by an external electric field acting on an atom schematized as a two-level system, with energy states $|a\rangle$ and $|b\rangle$, and corresponding eigenvalues E_a and E_b. Suppose that a time-dependent external electric field at angular frequency ω and amplitude E_0 is

5.2 Atomic Interferometers

acting on the atom, the corresponding potential energy term in the time-dependent Schrödinger equation will be

$$\hat{V}(t) = \mathbf{p} \cdot \mathbf{E}_0 \cos(\omega t)|a\rangle\langle b|, \tag{5.10}$$

with $\mathbf{p} = e\mathbf{r}$ the electric dipole moment. This potential energy term can be considered as a perturbation to the original two-level system dynamics, and its effect can be incorporated in the time dependence of the coefficients $c_a(0) = \langle a|\psi(0)\rangle$ and $c_b(0) = \langle b|\psi(0)\rangle$ of the expansion of the generic initial state $|\psi\rangle(0)$, such that at generic time the state will be written as

$$|\psi(t)\rangle = c_a(t)e^{-iE_a t/\hbar}|a\rangle + c_b(t)e^{-iE_b t/\hbar}|b\rangle. \tag{5.11}$$

Inserting this expression into the Schrödinger equation, one obtains equations for the coefficients $c_a(t)$ and $c_b(t)$ one obtains

$$i\hbar\dot{c}_a(t) = \Omega \cos(\omega t)e^{+i(E_a-E_b)t/\hbar}, \tag{5.12}$$

$$i\hbar\dot{c}_b(t) = \Omega \cos(\omega t)e^{-i(E_a-E_b)t/\hbar}, \tag{5.13}$$

where we have introduced the Rabi frequency $\Omega_R = \langle a|\mathbf{p}\cdot\mathbf{E_0}|b\rangle/\hbar$. It is also useful to introduce $\omega_{ab} = (E_a - E_b)/\hbar$, the transition angular frequency between the two levels. Therefore, we have three distinct angular frequencies, ω_{ab} intrinsic to the two-level system, ω which determines the timescale of the external driving due to the electric field, and Ω_R which depends on both the atom and the external electric field. In the case we are considering, ω_{ab} and ω are in the visible range of the electromagnetic spectrum; therefore, $\omega_{ab} \simeq \omega$. Under this assumption, one can regroup the terms in Eq. (5.13), using the exponential form of the cosine, as the same of terms oscillating at $\omega_{ab} + \omega$ and $\omega_{ab} - \omega$, and neglect the first term with high-frequency oscillations at $\simeq \omega_{ab}$, the so-called rotating-wave approximation. Then, Eq. (5.13) can be solved analytically, and starting, for instance, from an initial condition $c_a(0) = 1$, we obtain the probability to observe the system in state $|b\rangle$ at later time t as

$$P_b(t) = |c_b(t)|^2 = \frac{\Omega_R^2}{\Omega_R^2 + (\omega_{ab} - \omega)^2} \sin^2\left(\frac{\sqrt{\Omega_R^2 + (\omega_{ab} - \omega)^2}}{2} t\right). \tag{5.14}$$

This shows that a complete population transfer is obtained if $\omega = \omega_{ab}$ (i.e., if the photons are sent in resonance with the two-level system) and the driving of the atoms ends at a time equal to $T_\pi = \pi/\Omega_R$. This realizes a so-called π-pulse. If instead one stops the laser beam at a time equal to $T_{\pi/2} = T_\pi/2$, the system will be in a superposition state of $|a\rangle$ and $|b\rangle$ with equal coefficients, a $\pi/2$ pulse. This last can be considered a protocol to create a beam splitter for atoms. Notice that laser beams

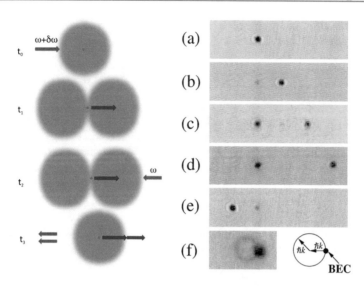

Fig. 5.3 (left) Schematics of Bragg scattering using two counterpropagating laser beams. Unlike the case of spontaneous emission, see Fig. 2.3, in this case the entropy is conserved during the process; initial and final states of atoms and photons are simply rearranged with changes in momentum without any intrinsic randomness. (right) Absorption images taken 5.6 ms after a Bragg pulse with different detunings $\delta\omega$, corresponding to values of 0 in (**a**) (condensate unaffected), 98 kHz in (**b**) (almost complete conversion into a state with one recoil unit, initial condensate barely visible), 200 kHz in (**c**) (two recoil units, $\pi/2$ pulse to split the condensate into two equally populated modes), 300 kHz in (**d**) (three recoil units, π pulse), and −98 kHz (**e**) (recoil mode as in (**b**) but in opposite direction). The bottom image (**f**) shows the effect of a single pulse, absorption process with a recoil, followed by a recoil in a random direction, forming a halo of atoms lying on a circle, the distinctive feature of spontaneous emission discussed in Chap. 2. Credit: (Right) reprinted figure with permission from [29]. Copyright 1999 by the American Physical Society

with different pulse durations and different intensities can result in the same π (or $\pi/2$) pulse provided that $T_{\pi/2} E_0$ is constant.

In the simplest case of two counterpropagating light beams, as in the left panel of Fig. 5.3, the beams have a relative frequency detuning $\delta\omega$ and are successively sent on the condensate at rest in the trap, resulting in momentum release to the atoms with controlled direction. The atom absorbs a photon with energy $\hbar(\omega + \delta\omega)$ from the beam impinging from its left, resulting in a first recoil of amplitude $\hbar k$ in the horizontal direction. The excited atom undergoes a stimulated emission process due to the beam impinging from the right, emitting a photon identical to the one stimulating the process, with energy $\hbar\omega$ and resulting in a second recoil of amplitude $\hbar k$ in the same direction as the first one.

More in general, if the generic angle between the two beams is θ not necessarily equal to π, a process of recoil will consist of absorption of one photon from the beam with higher frequency and stimulated emission into the beam with lower frequency, with a momentum recoil of $p_r = 2\hbar k \sin(\theta/2)$. Conservation of energy imposes that the kinetic energy gained by the atoms in the recoil process $p_r^2/(2m)$, with m

5.2 Atomic Interferometers

the mass of the atom, equals the energy exchanged with the laser beams, $\hbar\delta\omega$. The Bragg resonance condition for noninteracting atoms if multiple photons are involved in the process, for instance, N photons are absorbed and N photons are emitted in a stimulated process, results in a Bragg scattering process of Nth order. The energy balance for this generic process yields

$$\frac{(Np_r)^2}{2m} = N\hbar\delta\omega, \tag{5.15}$$

showing that choosing integer multiples of $\delta\omega$ allows to select a targeted recoil mode of order N. Intensity and duration of the laser pulses control the conversion of a defined amount of atoms from the condensate into the desired recoil mode, for instance, realizing a π-pulse in the case of complete conversion or a $\pi/2$ pulse for a 50-50 splitting. Bragg scattering has been demonstrated for Bose-Einstein condensates at NIST Gaithersburg [29]; see panels on the right of Fig. 5.3. A Bose-Einstein condensate of ^{23}Na atoms was adiabatically expanded, to reduce its momentum spreading due to in situ mean-field energy, and then illuminated with laser pulses of assigned frequency detuning. As visible in the various panels, efficient conversion of nearly the whole condensate into states with defined momentum in integer multiples of the recoil velocity was achieved. This is the basis for realizing beam spitters and mirrors in an atomic interferometer. Although the experiment described in [29] allows to visualize Bragg scattering, the process was already studied with atomic beams earlier [35] and thereafter applied to atomic interferometers with thermal sources [19].

For an atomic 50-50 beam splitter, its output will be written as

$$|\psi_{BS}\rangle = \frac{1}{\sqrt{(N+1)!}}(\hat{a}^+ + \hat{b}^+)^N|0\rangle, \tag{5.16}$$

with N the number of atoms impinging on the beam splitter and $\hat{a}^+$, $\hat{b}^+$ the creation operators in states $|\psi_a\rangle$, $|\psi_b\rangle$, respectively. These states can differ by external observables, such as the direction of momentum, or internal observables, internal atomic states, as in Ramsey interferometry. The presence of an external agent acting in different ways on the two modes, such as gravity or a medium, will induce a relative phase shift ϕ between the two modes, and then, the state will change into $|\psi_{BS}\rangle \propto (\hat{a}^+ + \exp(i\phi)\hat{b}^+)^N|0\rangle$. This allows to calculate the average number of atoms in modes a and b once recombined at the detector, an atom counter, and the quantity of interest is $n = N_A - N_B$, which is a sinusoidal function of ϕ. The minimum detectable phase is determined by the sensitivity of ϕ to the measured number of atoms n, as $\Delta\phi = \Delta n/(\partial n/\partial \phi)$. The denominator is minimized by working at the point of maximum slope in the $n(\phi)$ dependence and in this case is the fringe visibility, supposed to be close to unity for a quasi-ideal setup, while Δn depends on the specific preparation of $|\psi_{BS}\rangle$. In a "standard" situation, without special precautions, Δn for an N-atom state is calculable by considering the uncertainty for N independent measurements on a single atom, such that $\Delta\phi \simeq 1/\sqrt{N}$. This is also called "standard quantum limit," as consequence of the number-phase uncertainty relationship for which $\Delta N \, \Delta\phi = 1$, as discussed in the former section. This shows the relevance of sending high-brightness fluxes of atoms once technical sources of noise, in particular those limiting the coherence of the beam, are properly mitigated.

We also briefly discuss the case of interferometry with ultracold fermions. Using again the second quantization approach, and the anticommutation rules for fermions, in this case, the only possibility is to send one atom on the beam splitter, to achieve the quantum state

$$|\psi_{BS}\rangle = \frac{1}{\sqrt{2}}(|1\rangle \otimes |0\rangle - |0\rangle \otimes |1\rangle), \qquad (5.17)$$

and $\langle N_a \rangle = \langle N_b \rangle = 1/2$, $\langle N_a^2 \rangle = \langle N_b^2 \rangle = 1/2$; therefore, the standard deviations are $\Delta N_a = \Delta N_b = 1/2$. The resulting large variance precludes the possibility of measuring phase shifts with high precision, and fermion atomic interferometry, as also for electron and neutron interferometry, is only possible in the low brightness, nondegenerate regime. This is regrettable, as fermions in the degenerate regime act as a nearly collisionless gas, potentially giving rise to long coherence times. The potentiality of fermions in this regard has been demonstrated in [14] by studying Bragg scattering on an ultracold cloud of ^{6}Li atoms.

Among the various implementations of atomic interferometers, particularly advantageous are the ones in gravimetry. By considering an interferometer with the two paths lying along a vertical plane, there will be a gravitational phase shift between the upper path and the lower path equal to $\Delta\phi_{\text{atom}} = mgL\Delta h/(\hbar v)$ if the horizontal extension of the two paths has length L (31 cm in the example reported in Fig. 5.2), and they are spaced by Δh in the vertical direction, with v the atom velocity. For a optical Mach-Zehnder interferometer, we expect $v = c$, $m = \hbar\omega/c^2$, and therefore, $\Delta\phi_{\text{photon}} = \omega g L \Delta/c^3$. The ratio of the phase shifts is then

$$\frac{\Delta\phi_{\text{atom}}}{\Delta\phi_{\text{photon}}} = \frac{mc^2}{\hbar\omega} \frac{c}{v}. \qquad (5.18)$$

The first ratio in the right-hand side is quite large for atoms of atomic mass A having rest energy $\simeq A \times 1\,\text{GeV}$ and photons in the visible $\simeq 1 - 10\,\text{eV}$. Even more gain is expected from the second ratio appearing in the right-hand side, since atoms can be quite nonrelativistic, for instance, with velocities of order of cm/s if imparted by photon recoil. However, for the same reason of having enhanced sensitivity, atoms are also more prone to decoherence; therefore, they cannot match the same extension of the photons paths, both in L and Δh. In addition, atomic sources do not have the same brightness of laser sources for light interferometers; therefore, the detection count has larger statistical errors. Nevertheless, the progress in gravimetry using thermal sources of atoms has been enormous, with precise measurements of the gravitational acceleration [41], the measurement of the Newtonian universal constant of gravitation [5, 17, 31], and even tests of the equivalence principle with atoms as test masses [18], including its possible quantum formulation [44]. Although the precision of some of these measurements is not yet competitive with the one obtained from macroscopic test masses, such as Eotvos balances in the case of tests of the equivalence principle, there are still margins of improvement. The systematic sources of error in atomic gravimetry are completely different from the ones present in more traditional macroscopic apparatus, on top of the advantages of using universal, indistinguishable systems provided by nature instead of fabricated ones. In light of this framework, it is therefore natural to discuss the possible application of ultracold atoms as atomic sources for interferometry, thereby extending the notion of optical lasers to the atomic realm. Although we will see that more progress is still required for implementing a competitive matter-wave interferometer using ultracold atoms, the possibility to increase the sensitivity through collective effects such as superradiance and entanglement, already interesting topics in themselves, provides reasons for optimism.

5.3 Atom Lasers

A crucial property of a Bose condensate is the existence of a macroscopic wave function, including a common phase, or a spatially variable phase with a predictable pattern. This phase is crucial for the description of the transport properties of the condensate, such as superfluidity, soliton, and vortex formation, but it is also important to assess the sensitivity for interferometry based on the use of coherent atomic beams, the so-called atom lasers. This opens the possibility to study interference from two different atom sources, in analogy to the case of optical interferometers. Coherence of atomic samples was initially demonstrated at MIT by interfering two Bose condensates in free fall, as described in Chap. 3. Based on the significant visibility in the observed interference fringes even after the free fall of matter waves, and the physical meaning of the macroscopic wave function, it is expected that spatial coherence extends to the whole condensate, at least as long as thermal effects are not dominant. Furthermore, it is possible to quantitatively confirm this expectation by performing Bragg spectroscopy [47]. The idea is to measure the momentum distribution of a trapped Bose-Einstein condensate by absorption of photons from one laser beam and successive stimulated emission by another laser beam, as discussed earlier. The main advantage of Bragg spectroscopy is that it is a stimulated, background-free process in which the transfer of energy and momentum to the atoms is predetermined by the frequency and the geometry of the laser beams, instead of being postdetermined from measurements of energy and momentum of the atoms. The momentum distribution of the atoms will be inferred by scanning the frequency detuning between the two beams and measuring the number of atoms ejected from the condensate. Such a momentum distribution is the square modulus of the wave function in momentum representation, in turn related via a Fourier transform to the wave function in position representation. In the Thomas-Fermi approximation, the rms width of the momentum distribution along the x axis in which the two laser beams lie is related to the condensate size x_0 as $\Delta p_x = \sqrt{21/8}\hbar/x_0$, related to the uncertainty principle for position and momentum. The experiment tested the dependence of the momentum distribution upon the density of the condensate and also investigated its dependence on the mean-field energy, resulting in both an energy shift and a broadening of the resonance curve (Fig. 5.4).

Further experiments were carried out in Munich, exploiting the high stability of the magnetic trap, and allowing for a quantitative study of the fringe visibility dependence on crucial parameters such as the temperature of the condensate [6]. The configuration combined quadrupole and Ioffe-Pritchard geometries in a compact setup, with magnetic field fluctuations due to the environment reduced below 0.1 mG. The atoms are expelled from the magnetic trap by means of rf-coupling at two defined angular frequencies ω and ω'. The two outgoing matter waves then have an energy difference of $\Delta E = \hbar|\omega - \omega'|$, and this energy is chosen to reflect a spatial separation along the vertical $\Delta z = \Delta E/(mg)$ equal to the turning points of the solutions, expressed in terms of Airy functions, for the one-dimensional Schrödinger problem of the outcoupled wave in the presence of uniform gravity. The matter waves are extracted from the condensate over a period of 13 ms. After a delay of 2 ms, the magnetic field is switched off and an absorption image is taken after further 3 ms.

The fringe visibility is obtained by fitting the absorption image with the atomic density distribution expected for two output waves originating from two points vertically spaced by Δz, giving rise to interference

$$n_{\text{out}} = |\psi_{\text{out}}(\zeta) + \psi_{\text{out}}(\zeta')|^2 \propto \frac{1}{\sqrt{z}} \left\{ 1 + V \cos\left[q\sqrt{z} + (\omega - \omega')t\right] \right\}, \quad (5.19)$$

where $0 \leq V \leq 1$ is the fringe visibility. In Fig. 5.5, the dependence of the resulting fringe visibility is shown versus the temperature, including data taken above the critical temperature for Bose-Einstein condensation T_c. In the latter case, the coherence length of the cloud is smaller than the slit separation of 325 mm. Below T_c, a sudden increase in visibility is observed, reaching almost unity at the lowest explored temperatures. Notice that a fringe visibility of about 0.7 is maintained at the largest explored distance between the two point sources, of 700 nm, in the case of $T = 250$ nK. The temperature was determined from a bimodal fit to the absorption images of the condensate, with a 20 nK precision, with a critical temperature for Bose-Einstein condensation $T_c = 430$ nK.

Other experiments on the coherence of outcoupled matter waves have been performed at Yale University by using a 1D optical lattice vertically superimposed to the magnetic trap holding the condensate [3]. The sinusoidal effective potential created by the optical lattice is modulated by the gravitational potential; therefore, the energy difference between two contiguous wells is $mg\lambda/2$, with λ the wavelength of the (red-detuned) laser beam. Each lattice site can be modelled as a point emitter of de Broglie waves with an emission rate proportional to the tunneling probability. In subsequent experiments, the same group has created squeezed states of atoms

5.3 Atom Lasers

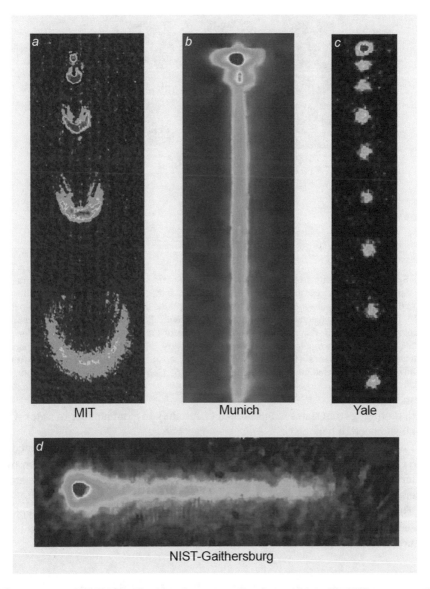

Fig. 5.4 Examples of early realizations of output coupling for atom lasers. The MIT group created a pulsed atom laser based on the ejection of antitrapped atoms with the application of spin-flip rf-pulses equally spaced in time. Notice the "crescent moon" shape of each atomic cloud in free fall, due to their strong interatomic repulsion. In Munich, a similar output coupling was adopted but using two different radiofrequencies spilling over antitrapped atoms from two definite locations inside the condensate and creating interference patterns not visible here (instead appearing in Fig. 5.5). At Yale, the atoms were emitted by an array of sources created by a vertical optical lattice superimposed to the trapping potential, resulting in their interference pattern during the free fall. At NIST Gaithersburg, the output coupler was based on stimulated Raman transitions, imparting momentum in an arbitrary direction, for instance, horizontal in this image. Credit: reprinted figure from [23], NIST

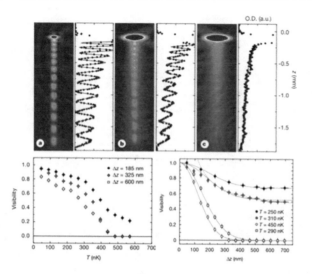

Fig. 5.5 Fringe visibility from two point sources of atomic waves in the presence of gravity. On top, absorption images (field of view of 0.6 mm × 2 mm) of the matter waves emitted from two different locations inside a stable magnetic trap, for three different temperatures, well below (left), close and below (center), and close and above (right) T_c. In this last situation, coherence is completely lost, indicating that the spatial coherence of the matter wave is smaller than the distance between the two point sources. More quantitatively, in the bottom plots, the dependence on temperature at constant slit separation (left) and on slit separation at constant temperature (right) is shown. Credit: reprinted figure with permission from [6]. Copyright 2000 by Springer Nature

[39], morphing from states with minimal uncertainty in atom number to states with minimal uncertainty in phase.

Finally, an output coupler with arbitrary directionality has been demonstrated at NIST, Gaithersburg, by means of absorption and stimulated emission between two different energy levels, realizing Raman scattering [22]. A panoply of these atom lasers is depicted in Fig. 5.4.

The analogous of a continuous-wave optical laser requires a continuous replenishment of the reservoir of atoms in the Bose-Einstein condensate, and this has been demonstrated in [9, 10].

5.4 Superradiance and Matter-Wave Amplification

Further intriguing phenomena have been studied in connection to the response of a Bose-Einstein condensate to spontaneous emission events. This has led to the identification of a mechanism to amplify matter waves in a well-defined mode and preserving the phase, an important step toward the matching of atom lasers to atomic interferometers. The starting point is Rayleigh scattering, in which absorption of a photon is then followed by spontaneous emission. The first event is related to the direction of the impinging photon, while the second is random,

5.4 Superradiance and Matter-Wave Amplification

a fact already exploited for slowing an atomic beam as described in Chap. 2. If the atoms are immersed in a Bose-Einstein condensate, we have to keep in mind that long coherence times are available. If these are longer than the average time interval between two scattering events, the emission of photons will no longer be uncorrelated. The scattering of the first photon leaves an imprint in the condensate density profile that survives until a second photon is absorbed. But this second photon will not encounter the same condensate profile as the first, and the directional probability will be accordingly changed. This keeps going providing directional Rayleigh scattering, a more organized form of spontaneous emission. A semiclassical model allows to grasp the dynamics of the process. When a condensate of N_0 atoms is exposed to a laser beam with wave vector $\mathbf{k_0}$ and emits a photon with wave vector $\mathbf{k_j}$, the atom will get a recoil momentum $\hbar \mathbf{k_j} = \hbar(\mathbf{k_0} - \mathbf{k_i})$. Since the light propagates at a velocity about 10 orders of magnitude greater than the atomic recoil velocity (order of 3 cm/s in sodium), the recoiling atoms remain within the volume of the condensate long after the photons have left. These atoms will then interfere with the condensate at rest forming a matter-wave grating of wave vector $\mathbf{k_j}$, available to diffract incoming photons with the phase-matching direction $\mathbf{k_i} = \mathbf{k_0} - \mathbf{k_j}$. This is a self-amplifying process as each diffracted photon creates another recoiling atom in turn increasing the contrast in the matter-wave grating. If N_j is the number of recoiling atoms interfering with N_0 atoms at rest, the density modulation will be $N_{\text{mod}} = 2\sqrt{N_0 N_j}$, and the light scattered in the phase-matching direction will have a total power:

$$P = \frac{1}{4} f_j \hbar \omega R N_{\text{mod}}, \tag{5.20}$$

where R is the single-atom Rayleigh scattering rate, ω the angular frequency of the photons, f_j a mode density factor $f_j = 3 \sin^2 \theta_j \Omega_j /(8\pi)$, with θ_j the angle between the polarization of the incident light and the direction of emission, and $\Omega_j \simeq \lambda^2/A$ is the sold angle over which phase-matching is ensured, with A the cross-sectional area of the condensate perpendicular to the direction of emission of the light, and λ the wavelength of the photons. This matching is optimal for emission of light along the axis in which the condensate is more elongated, also called end-fire mode (Fig. 5.6).

Since each scattered photon creates a recoiling atom, their growth rate is

$$\dot{N}_j = \frac{P}{\hbar \omega} = R f_j N_0 N_j, \tag{5.21}$$

indicating exponential growth with the gain $R f_j N_0$, at least initially. This model is valid for $N_j \ll N_0$, in absence of decoherence, and indicates the buildup of anisotropic Rayleigh scattering from nonspherical Bose-Einstein condensates, depending upon the polarization state of the light with respect to the elongated cloud. Experiments were conducted at MIT and at Tokyo University. At MIT, an elongated condensate was exposed to a single off-resonant laser pulse detuned by

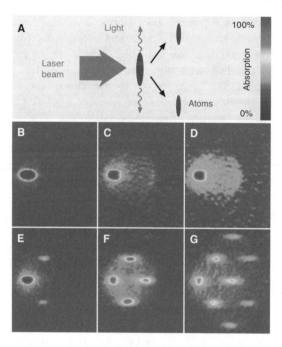

Fig. 5.6 Superradiance in a Bose-Einstein condensate. A laser pulse imparts momentum and energy to a few atoms in the condensate, and they then spontaneously emit a photon adding energy and momentum due to their recoil. Absorption images are then taken after a time of flight of 20 ms. The three columns are distinct by the duration of the laser pulse, increasing from left to right. If the impinging laser pulse has a polarization along the long axis of the condensate, usual Rayleigh scattering occurs, with the spontaneous emission as expected for isolated atoms (see absorption images (**b**) to (**d**)), with a dipole emission pattern. If instead the polarization of the laser pulse is orthogonal to the long axis spontaneous emission is a collective effect, and photons are emitted mainly along the symmetry axis of the condensate, in the "end-fire" modes (see absorption images (**e**) to (**g**)). This originates directional emission of atoms in a cascade due to their multiple interference. Credit: reprinted figure with permission from [25]. Copyright 1999 by the American Association for the Advancement of Science

1.7 GHz from the $3s_{1/2}(F = 1) \rightarrow 3p_{3/2}F = 0, 1, 2)$ transition. The beam was sent on the elongated condensate perpendicular to its long axis, with variable laser intensities corresponding to Rayleigh rates between 45 Hz and 4.5 kHz, and pulse duration between 10 and 800 μs. Absorption images were then taken, showing the momentum distribution of the scattered atoms. For polarization parallel to the long axis ($\theta_i = 0$), the distribution of the recoiled atoms followed the dipolar pattern of usual, single-atom, Rayleigh scattering. For orthogonal polarization ($\theta = \pi/2$), photons were emitted mainly along the long axis of the condensate, with the recoiling atoms appearing as clusters propagating at $\pi/4$ with respect to the condensate long axis. By repeating the measurements at variable laser intensities, a

5.5 Josephson-like Interferometry

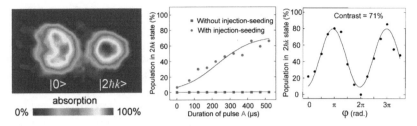

Fig. 5.7 Coherent amplification of matter waves. (Left) absorption image taken after a time of flight of 15 ms after a Bragg pulse (field of view 400 μm × by 260 μm). (Center) number of atoms in the $2\hbar k$ momentum state versus the duration of the superradiant pulse with and without injection of matter waves as input state of the amplification process. (Right) test of the phase-sensitive amplification, with use of a Mach-Zehnder interferometer to determine the matter-wave population in the $2\hbar k$ momentum state versus the phase of the injected matter wave. Credit: reprinted figure with permission from [30]. Copyright 1999 by the American Physical Society

threshold for superradiance was evidenced. This can be accounted for by adding a loss term in the growth equation:

$$\dot{N}_j = (G_j - L_j)N_j, \tag{5.22}$$

and fitting the initial rise in the scattered light intensity along the end-fire mode. From this fit, a decoherence time $L_j = 35\,\mu s$ was determined. Finally, a test of the coherence of the amplification process has been performed, as shown in Fig. 5.7, by using a Mach-Zehnder interferometer. The Tokyo group measured a visibility contrast of 71 % with up to 80 % of the atoms transferred from the condensate at rest into a state with momentum $2\hbar k$.

The use of condensates as initial states for atomic interferometers is limited by the strong mean-field energy of the atoms, the reason for which in the experiment reported in [29] the condensate was adiabatically expanded prior to the Bragg pulse, although this issue can be circumvented by changing the scattering length. As long as one uses atoms in defined momentum states, in "flight," there is no substantial advantage in using Bose-Einstein condensates. As we will discuss at the end of the next section, it is however possible to counterbalance this drawback by exploiting the quantum correlation present in a pure Bose-Einstein condensate, creating entangled states.

5.5 Josephson-like Interferometry

Not all interferometers are based on flying particles in momentum eigenstates. Already at the optical level, photons can be trapped in cavities, with the most known example of the Fabry-Perot interferometer. In analogy, atoms can be trapped, and their dynamics can be chosen in such a way that there are at least two alternative paths that can then recombine together after many cycles, thereby enhancing the potential phase difference due to an external agent. The minimal requirement is a potential, at least in one dimension, which is bistable, with two local minima, among which the atoms can hop. In the case of free particles, the preparation of a momentum state at a given time does not perturb its successive values as momentum is conserved in the absence of an external potential. Instead, if a particle is prepared in a position state in an external potential, the uncertainty resulting in the momentum at the time of preparation will feed back on the position uncertainty at later time, originating so-called back action. Care must be exerted then; in the full quantum regime, we want to consider these class of interferometers, when measuring the system after a fixed evolution time. To visualize the dynamics, minimally invasive measurements should be used, for instance, phase-contrast imaging. However, even in this case the measurement process will potentially scramble the phase, generating decoherence [12]. In alternative, a single-shot measurement of the resulting fringes at the end of the dynamics should be performed. This suggests the description of the system in terms of atom number and phase for the two paths, possibly including their fluctuations. It is then natural to use the second quantization approach described in Appendix A. Moreover, in this way, the discussion will be suitable to deal with ultracold bosons and fermions alike, although for reasons of space we will limit the analysis to the former case.

As discussed for the reduction to the one-dimensional case in Chap. 3 in regard to the formation of solitons, we assume a trapping potential as in Eq. (3.28), with the specific choice of $\mathcal{U}(z) = V_{DW}(x)$, a double-well potential along the z axial direction. The double-well potential can be achieved by a proper configuration of the magnetic trap or by the use of a time-averaged optical potential with a red-detuned laser beam quickly alternating the focus, using an acousto-optical modulator, between two points lying on the symmetry axis and located at the same distance from the center of the trap.

Under the assumption of large angular frequency $\omega_\perp$, the dynamics along the transverse confinement in the x and y plane is frozen, and the solution of the many-body equation for the bosonic field operator is approximated as

$$\hat{\psi}(\mathbf{r}) = \hat{\phi}(z) \frac{e^{-(x^2+y^2)/(2l_\perp^2)}}{\pi^{1/2} l_\perp}, \qquad (5.23)$$

where $a_\perp = \sqrt{\hbar/(m\omega_\perp)}$ is the characteristic length of the transverse confinement (Eq. (3.29)). Inserting Eq. (5.23) into the Hamiltonian Eq. (A.47) and integrating over yx and y, we obtain the effective 1D Hamiltonian:

$$\hat{H} = \int dz \hat{\phi}^+(z) \left[-\frac{\hbar^2}{2m} \frac{d^2}{dz^2} + V_{DW}(z) + \hbar\omega_\perp \right] \hat{\phi}(z)$$
$$+ \frac{g_{1D}}{2} \int dz \, \hat{\phi}^+(z)\hat{\phi}^+(z)\hat{\phi}(z)\hat{\phi}(z), \qquad (5.24)$$

5.5 Josephson-like Interferometry

with g_{1D} the effective 1D interaction strength defined after Eq. (3.30). We suppose that the barrier of the double-well potential $V_{DW}(z)$, with its local maximum located at $z = 0$, is high enough to sustain several doublets of quasi-degenerate single-particle energy levels. Moreover, we suppose that only the lowest doublet (i.e., the single-particle ground-state and the single-particle first excited state) significantly contributes to the state, a two-mode approximation. Under these assumptions, we can write the bosonic field operator as

$$\hat{\phi}(z) = \hat{a}_L \, \phi_L(z) + \hat{a}_R \, \phi_R(z) \tag{5.25}$$

that is the so-called two-mode approximation, where $\phi_L(z)$ and $\phi_R(z)$ are single-particle wave functions localized, respectively, on the left well and on the right well of the potential $V_{DW}(z)$. These wave functions are linear combinations with equal coefficients of the even wave function $\phi_0(x)$ of the ground state and the odd wave function $\phi_1(x)$ of the first excited state. The operator $\hat{a}_j$ annihilates a boson in the j-th site (well), while the operator $\hat{a}_j^+$ creates a boson in the j-th site ($j = L, R$).

Inserting the two-mode approximation (5.25) of the bosonic field operator in the effective 1D Hamiltonian, we get the following two-site Hamiltonian:

$$\hat{H} = \epsilon_L \hat{N}_L + \epsilon_R \hat{N}_R - J_{LR} \hat{a}_L^+ \hat{a}_R - J_{RL} \hat{a}_R^+ \hat{a}_L + \frac{U_L}{2} \hat{N}_L(\hat{N}_L - 1)$$
$$+ \frac{U_R}{2} \hat{N}_R(\hat{N}_R - 1) , \tag{5.26}$$

where $\hat{N}_j = \hat{a}_j^+ \hat{a}_j$ is the number operator of the j-th site

$$\epsilon_j = \int dz \, \phi_j(z) \left[-\frac{\hbar^2}{2m} \frac{d^2}{dz^2} + V_{DW}(z) + \hbar \omega_\perp \right] \phi_j(x) , \tag{5.27}$$

is the kinetic plus potential energy on the site j,

$$J_{ij} = \int dz \, \phi_i(z) \left[-\frac{\hbar^2}{2m} \frac{d^2}{dz^2} + V_{DW}(z) + \hbar \omega_\perp \right] \phi_j(z) , \tag{5.28}$$

is the hopping energy (tunneling energy) between the site i and the site j, and

$$U_j = g_{1D} \int dz \, \phi_j(z)^4 , \tag{5.29}$$

is the interaction energy on the site j.

The Hamiltonian (5.26) is the two-site Bose-Hubbard Hamiltonian, named after John Hubbard introduced a similar model in 1963 to describe fermions on a periodic lattice, in which case j is not limited to two entries, but runs over M sites

$$\hat{H} = \epsilon \sum_{j=1}^{M} \hat{N}_j - J \sum_{j=1}^{M-1} \left(\hat{a}_j^+ \hat{a}_{j+1} + \hat{a}_{j+1}^+ \hat{a}_j \right) + \frac{U}{2} \sum_{j=1}^{M} \hat{N}_j (\hat{N}_j - 1) . \quad (5.30)$$

If the double-well potential $V_{DW}(z)$ is fully symmetric, then $\epsilon_L = \epsilon_R = \epsilon$, $J_{LR} = J_{RL} = J$, $U_L = U_R = U$, and the Bose-Hubbard Hamiltonian becomes

$$\hat{H} = \epsilon \left(\hat{N}_L + \hat{N}_R \right) - J \left(\hat{a}_L^+ \hat{a}_R + \hat{a}_R^+ \hat{a}_L \right) + \frac{U}{2} \left[\hat{N}_L (\hat{N}_L - 1) + \hat{N}_R (\hat{N}_R - 1) \right] . \quad (5.31)$$

The Heisenberg equations of motion of the operator $\hat{a}_j$ ($j = L, R$) are

$$i\hbar \frac{d}{dt} \hat{a}_j = -J \hat{a}_i + U \hat{N}_j \hat{a}_j . \quad (5.32)$$

By averaging the Heisenberg equation of motion on coherent states, we find

$$i\hbar \frac{d}{dt} \alpha_j = -J \alpha_i + U |\alpha_j|^2 \alpha_j , \quad (5.33)$$

where $\alpha_j(t) = \bar{N}_j(t)^{1/2} e^{i\theta_j(t)}$ with $\bar{N}_j(t)$ the average number of bosons in the site j at time t and $\theta_j(t)$ the corresponding phase. Working with a fixed number of bosons, i.e., $N = \bar{N}_L(t) + \bar{N}_R(t)$ and introducing two new parameters, the population imbalance and the relative phase, respectively, as

$$z(t) = \frac{\bar{N}_L(t) - \bar{N}_R(t)}{N}, \quad \theta(t) = \theta_R(t) - \theta_L(t) , \quad (5.34)$$

the time-dependent equations for $\alpha_L(t)$ and $\alpha_R(t)$ can be rewritten as follows:

$$\frac{dz}{dt} = -\frac{2J}{\hbar} \sqrt{1 - z^2} \sin(\theta) , \quad \frac{d\theta}{dt} = \frac{2J}{\hbar} \frac{z}{\sqrt{1 - z^2}} \cos(\theta) + \frac{UN}{\hbar} z . \quad (5.35)$$

These equations describe the dynamics of the population imbalance $z(t)$ and relative phase $\theta(t)$ of identical bosons undergoing macroscopic quantum tunneling. Under the condition of small population imbalance ($|z| \ll 1$), one finds

$$\frac{dz}{dt} = -\frac{2J}{\hbar} \sin \theta , \quad \frac{d\theta}{dt} = \frac{1}{\hbar} (2J \cos \theta + UN) z . \quad (5.36)$$

These equations were introduced by Brian Josephson to describe the electric current between two superconductors separated by a thin insulating barrier [28].

5.5 Josephson-like Interferometry

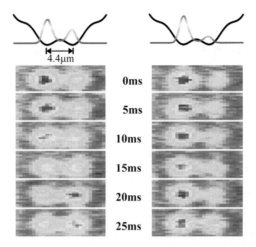

Fig. 5.8 Dynamics of a Bose-Einstein condensate in a double-well potential with minima at a distance of 4.4 μm. Absorption images with different evolution times are taken after increasing the distance between the minima to 6.7 μm for better spatial resolution. The time evolution is shown for two different values of the z parameter corresponding to tunneling oscillations (**a**) and macroscopic self-trapping (**b**). Credit: reprinted figure with permission from [1]. Copyright 2005 by the American Physical Society

In the limit of small population imbalance, the relative phase stays constant and consequently

$$\frac{dz(t)}{dt} = -\frac{2J}{\hbar} \sin\theta(0) \ . \tag{5.37}$$

This is the direct current (DC) Josephson effect: an initial phase difference $\theta(0)$ induces a constant current through the double-well barrier. As the population imbalance growths, if the relative phase $\theta(t)$ is still small, one finds

$$\frac{d^2z}{dt^2} + \left(\frac{2J}{\hbar}\right)^2 \left(1 + N\frac{U}{2J}\right) z = 0 \ . \tag{5.38}$$

This is the alternating current (AC) Josephson effect: there is a periodic oscillation of the population imbalance between the two wells of the double-well potential with angular frequency dependent on the number of atoms:

$$\Omega_J(N) = \frac{2J}{\hbar}\sqrt{1 + N\frac{U}{2J}} \ . \tag{5.39}$$

The complete dynamics, relaxing the $z \ll 1$ approximation, is indeed nonlinear in the number of atoms, and strong repulsive interatomic interactions can inhibit tunneling, a phenomenon called macroscopic self-trapping [46], demonstrated in [1] (see Fig. 5.8). Macroscopic quantum tunneling was first observed in supercon-

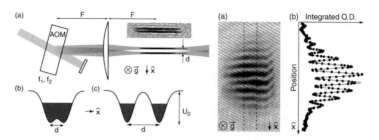

Fig. 5.9 Josephson interferometer with Bose-Einstein condensates. (Left) a double-well potential in an horizontal plane is generated with a red-detuned optical dipole trap and an acousto-optical modulator located in the focal plane of the lens focusing the laser beam and driven at two frequencies with a linear ramp of 5 ms duration, resulting in the time-averaged potential as in (d). (Right) absorption image after 30 ms time of flight (field of view 600 × 350 μm) and radial density profiles integrated over the dashed lines, with 60 % fringe visibility, with a best fit (continuous line) to determine the relative phase. Credit: reprinted figure with permission from [45]. Copyright 2005 by the American Physical Society

ducting junctions [2, 27], and it is at the basis of well-developed technology with impact in high-sensitivity magnetometry. Unlike their superconducting counterpart, the coherence of Josephson oscillations in the case of trapped ultracold atoms can be evidenced with interference patterns, as in Fig. 5.9. The preferential effect of an external agent on the phase of the condensate in one of the two wells can be enhanced by the cumulative effect due to many oscillations, as long as coherence can be maintained.

Interferometers based on localized states have the potential advantage of exploiting entanglement between the atoms to increase the precision beyond the classical limit. The limitation to the sensitivity of an atom interferometer is, provided more technical sources of noise are tamed, determined by the statistics in the visibility of the interference fringes, as we have discussed in Sect. 5.2 introducing the standard quantum limit. This holds if there is no correlation between the detected atoms, as in single-atom intererometry. For coherently correlated atoms, as in a Bose-Einstein condensate, using entangled states (see Appendix D), one could move the minimal phase error to the so-called Heisenberg limit, $\Delta\phi_{\text{Heis}} = 1/N$, a significant gain since $N \gg 1$ in an atomic (based on bosons) interferometer. A first demonstration of the gain achievable by correlating atoms has been described in [21]. The experiment uses a ^{87}Rb condensate in two hyperfine states, and a deep 1D optical lattice for which tunneling is negligible. The populations of the two hyperfine states are consecutively measured with absorption imaging. A nonlinear beam splitter is produced via Feshbach-induced strong interatomic interactions, and this creates coherent spin squeezed states starting from a state, created with a $\pi/2$ pulse, in which $\langle J_z \rangle = 0$. The phase uncertainty in terms of the spin of the atoms is $\Delta\phi = \xi_s/\sqrt{N}$ where ξ_s is an indicator of the degree of squeezing of one spin component, where

$$\xi_s = N\Delta J_z^2 / (\langle J_x^2 \rangle + \langle J_y^2 \rangle), \quad (5.40)$$

and the nonlinear time evolution transforms a "standard quantum limit" region of uncertainty with equal uncertainty into the two conjugate variables into a region in which the uncertainty is squeezed in one of the two variables. This then is converted into a state with squeezed phase uncertainty in J_z. The use of a squeezed state, with variance reduced by 8.2 decibels with respect to the one of the standard quantum limit, results in an estimated increase of 61 % in phase sensitivity. This looks quite promising, also considering that the estimated number of entangled atoms, about 170, is quite small with respect to what is currently targeted for scalable, massive quantum emulators (see also [24]).

5.6 Atomtronics and Quantum Emulators

The manipulation and control of macroscopic and microscopic degrees of freedom of atoms lend itself to the possibility to realize devices analogous to the ones already available and exploiting the manipulation and control of electric charge carriers such as electrons in conductors, electrons and holes in semiconductors, and Cooper pairs in superconductors. An example is already available with the experimental setups discussed in the previous section, mimicking the behavior of a Josephson junction. It should be remarked that using ultracold atoms, in the degenerate regime, results automatically in superfluid transport, the atomic counterpart of superconducting electronics, with appealing features especially as a platform for quantum computation and quantum emulation. Other counterparts of usual circuits in electronics can also be identified, leading to the possibility of an "atomtronics," as in the following examples:

- **Batteries.** An atomic battery should consist in two spatial locations with ultracold atoms maintained at different chemical potential. This can be achieved with two traps with different number of atoms and different trapping frequencies and a scheme to replenish the two reservoirs of atoms, for instance, with a continuous source [10].
- **Conductors and insulators.** An optical lattice can provide transport of atoms between two points at different chemical potential depending on the tunneling probability through each well, mimicking a conductor for high hopping probability (low potential depth of the lattice) or an insulator in the opposite case.
- **Waveguides.** This can be obtained with optical dipole traps with movable focus and results in traveled distances of tens of cm. The major issue is the need to avoid excitations of the ultracold atoms, which demands slow dynamics to achieve single-mode propagation. This is achieved, for instance, with time-averaged adiabatic potentials creating rings, with distances traveled of more

than 40 cm [40], or judiciously choosing the motion strategy, for instance, with shortcut to adiabaticity protocols [48].
- **Miniaturized magnetic traps.** These can be realized on semiconductor chips. Due to the smaller size, modest values of current densities are required for targeted magnetic field gradients and curvatures, and the wires can the cooled through the substrate. This allows to achieve trapping frequencies in the 10 KHz range and flexibility in designing complex patterns.
- **Doped materials.** These can be achieved by using lattices with two energy bands and putting in it atoms on the upper energy band, while the first band is filled (n-type lattice) or creating a depletion of atoms in the first energy band (p-type lattice). In this way, analogs of semiconductor p-n junctions and p-n-p or n-p-n junctions can be created, realizing diodes and transistors for atoms.
- **Amplifiers.** On top of in situ control of atomic flux with transistors as just discussed in the item above, one can use superradiance to amplify atomic signals; see Sect. 5.4.
- **Magnetic fields.** The so-called synthetic magnetic fields or, more in general, gauge fields can be simulated by imprinting vorticity to the atoms, especially using toroidal trap configurations, also including ring lattices.
- **Josephson junctions.** As described in the previous section, this is possible by using bistable potentials. A particularly promising configuration for weak links is provided by toroidal confinement, where Josephson tunneling has been demonstrated [42, 49].

An early example of an atomic circuit is shown in Fig. 5.10. Atoms are confined in a toroidal geometry by a combination of red-retuned light beams, an horizontal sheet created by focusing light with a cylindrical lens with focus on the center of

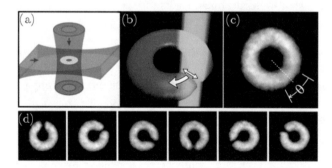

Fig. 5.10 Phase slip induced in a weak ring of a toroidal Bose-Einstein condensate. In (**a**) the geometry of the toroidal trap is shown, consisting of an optical dipole trap resulting from the intersection between a vertical ring-shaped beam and a horizontal sheet-shaped beam, both red detuned. (**b**) The barrier for the weak link is created with a blue-detuned beam quickly scanned in the radial direction with an acousto-optic deflector and azimuthally rotated at a lower frequency, up to 3 Hz. In situ absorption image of the atoms (field of view of 93 μm × 93 μm) in the ring without (**c**) and with (**d**) the barrier, shown after consecutive rotations by 60 degrees. Credit: reprinted figure with permission from [49]. Copyright 2013 by the American Physical Society

5.6 Atomtronics and Quantum Emulators

the toroid and a vertical ring beam. A blue-detuned light beam focused on the toroid is used to create a barrier weak link, and this beam is slowly rotated, creating a superfluid current and phase slips along the angle of the toroid, the closest analog to a single junction dc superconducting quantum interferometer device (SQUID). Increasing the rotation speed of the weak link resulted in the destruction of the phase slip and the formation of vortices entering the toroidal region. With respect to superconducting circuits or toroids of superfluid helium, an advantage is the possibility to control the critical current for formation of vortices by adjusting the barrier height.

It should be kept in mind that an atomtronic circuit has to rely on electronic circuits; therefore, nobody is expecting to replace the latter any time soon. It is however possible to foresee various "niche" applications in which they may result advantageous, starting from the fact that the transport is, provided the critical velocity is not exceeded, superfluid. After all, also superconducting electronics finds specific applications, but not a broad diffusion, also due to cost and complications of cryogenics devices. Another notable feature is that an atomtronic circuit can be programmed to perform several functions along the course of the run, for instance, a Mott insulator can be suddenly transformed into a superfluid lead. Components of atomic circuits on microchips, including the trapping and cooling setups, have been demonstrated, and further integration and scalability is within reach.

While the systematic implementation of circuits employing matter waves is a current frontier of quantum engineering with long-term impact hard to predict, ultracold atoms already have reached a degree of control such that models of quantum statistical mechanics can be emulated. The first step in this direction has been achieved by imaging of each individual site of an optical (single cubic) lattice with spacing of 4.9 μm created using blue-detuned light beams of wavelength 845.5 μm, with two beams in each direction (three orthogonal pairs) at a relative angle of 10 degrees [38]. Cesium atoms were loaded into the lattice after their trapping and cooling into a MOT followed by a stage of polarization-gradient cooling and imaged with a CCD camera detecting the scattered light. The key issue is to focus on only one plane of the lattice at any given time, and this is obtained by using a diffraction-limited objective outside the glass cell containing the atoms, which can be moved with a piezoelectric transducer to image other planes of the lattice.

The result is shown in Fig. 5.11 for three adjacent lattice planes (different columns) and two times since initial preparation, spaced by 3 seconds (different rows). The relatively large lattice spacing and potential depth prevent the observation of quantum dynamics such as hopping between sites due to quantum tunneling. However, at finite temperature, thermal hopping is observable, and the dependence of the hopping rate upon the potential depth was observed, in agreement with the Arrhenius law for thermal activation, allowing to measure the temperature of the atoms in the lattice. A second milestone was reported in [4], with progress in three aspects. First, the atoms, of ^{87}Rb, were brought to quantum degeneracy forming a 2D Bose-Einstein condensate trapped by both standing-wave and evanescent wave light fields. Second, a significantly smaller lattice spacing of 640 nm was achieved

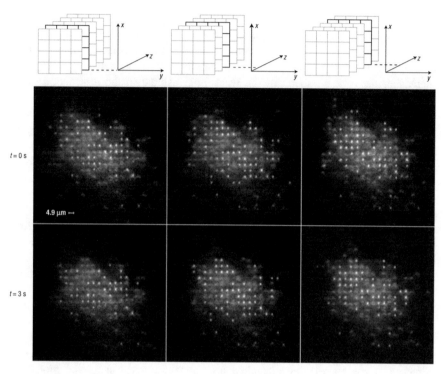

Fig. 5.11 Images from individual atoms in an optical lattice. The top three images are relative to three adjacent lattice planes in the x-y direction, as in the scketch above the images, with the imaging taking place along the z axis. The bottom three images are taken after 3 s from the former ones and evidence the dynamics due to both thermal hopping between different sites and loss of atoms due to background-gas collisions. Credit: reprinted figure with permission from [38]. Copyright 2007 by Springer Nature

by means of lithographically produced masks. Due to the smaller spacing, this lattice can sustain both hopping between adjacent sites through quantum tunneling and on-site confinement, having at the same time single site resolution imaging. Third, a high-resolution optical imaging system with resolution comparable to the lattice spacing was implemented, with a microscope objective at long working distance covering a large numerical aperture plus a hemispherical lens put inside the vacuum at a distance of a few micrometers from the condensate. This allows for high-fidelity imaging of the fluorescence from single atoms on single lattice sites.

These demonstrations allow for pursuing simulation of complex quantum systems, as envisaged by Feynman [16]. The simplest prototype could be an optical lattice with ultracold atoms in a Mott state, which will play the role of a memory in each lattice site. Quantum gates could be obtained with controllable interactions between atoms, with multiparticle, cluster-like systems in entangled states [34].

5.7 Problems

5.1. Calculate the phase shift induced in a Mach-Zehnder atom interferometer in which only one of the two paths is subjected to a uniform potential $U = 1\,\text{peV}$ for a length of 1 cm, using sodium atoms with a velocity of 10^3 m/s. Determine the effective refractive index corresponding to this potential.

5.2. Compare the momentum spread for the following initial states of an atomic interferometer using ^{87} atoms. (a) Atoms from an oven kept at $500\,^0$. (b) Cold atoms in a MOT at 1 mK. (c) 10^7 Ultracold atoms in a pure BEC state confined in an isotropic harmonic trap. Express the relative spread in units of momentum of a single recoil of resonant photon absorption.

5.3. The Heisenberg equation for a sodium atom moving with velocity is described by a density matrix in configurational space $\rho(x, x')$ can be modified to include the effect of decoherence of the environment as

$$\frac{\partial \rho(x, x')}{\partial t} = -\frac{i}{\hbar}[H, \rho(x, x')] - \Gamma \frac{(x-x')^2}{\lambda^2} \rho(x, x'),$$

where H is the Hamiltonian, Γ measure the rate of decoherence, and λ is a coherence length. Calculate the last two quantities for a sodium atom moving with velocity $v = 10^3$ m/s in an environment made of N_2 molecules in the residual pressure of $p = 10^{-10}$ mbar in a vacuum chamber kept at room temperature, assuming that it is enough one encounter of a sodium atom with a nitrogen molecule to disrupt its phase.

5.4. Extend the Josephson equations under the assumption that the double-well potential is not symmetric.

5.5. Determine the dependence of the period of oscillation of the nonlinear Josephson equations:

$$\ddot{\theta}(t) + \Omega^2 \sin \theta(t) = 0$$

upon the initial phase $\theta(0) = \theta_0$ and the number of atoms. Assume $J = 1$ and $U = 1$ in arbitrary units.

5.6. Two 3D optical lattices are formed with standing light waves at wavelengths of 2 μm and 500 nm, respectively. In both cases, the potential wells have a depth equivalent, in temperature, to 1 μK. Discuss if you can evidence, with an imaging system having spatial resolution of 400 nm, the presence of hopping between two adjacent sites by comparing two images, if the decoherence time is of 100 ms.

5.8 Further Reading

- M. Gross and S. Haroche, *Superradiance: An essay on the theory of collective spontaneous emission*, Phys. Rep. **93**, 301 (1982).
- P. Berman (editor), *Atom Interferometry* (Academic Press, San Diego, 1997).
- A. D. Cronin, J. Schmiedmayer, and D. E. Pritchard, *Optics and interferometry with atoms and molecules*, Rev. Mod. Phys. **81**, 1051 (2009).
- G. M. Tino and M. A. Kasevich (editors), *Atom Interferometry* (SIF and IOS Press, 2014).
- L. Pezzè, A. Smerzi, M. K. Oberthaler, R. Schmied, and P. Treutlein, *Quantum metrology with nonclassical states of atomic ensembles*, Rev. Mod. Phys. **90**, 035005 (2018).
- G.M. Tino, L. Cacciapuoti, S. Capozziello, G. Lambiase, and F. Sorrentino, *Precision gravity tests and the Einstein Equivalence Principle*, Progress in Particle and Nuclear Physics **112**, 103772 (2020).
- L. Amico, D. Anderson, M. Boshier, J.-P. Brantut, L.-C. Kwek, A. Minguzzi, and W. von Klitzing, *Colloquium: Atomtronic circuits: From many-body physics to quantum technologies*, Rev. Mod. Phys. **94**, 041001 (2022).

References

1. Albiez, M., Gati, R., Foelling, J.,Hunsmann, S., Cristiani, M., Oberthaler, M.K.: Direct observation of tunneling and nonlinear self-trapping in a single bosonic Josephson junction. Phys. Rev. Lett. **95**, 010402 (2005)
2. Anderson, P.W., Rowell, J.M.: Probable observation of the Josephson Superconducting tunneling effect. Phys. Rev. Lett. **10**, 230 (1963)
3. Anderson, B.P., Kasevich, M.A.: Macroscopic quantum interference from atomic tunnel arrays. Science **282**, 1686 (1998)
4. Bakr, W.S., Gillen, J.I., Peng, A., Folling, S., Greiner, M.: A quantum gas microscope for detecting single atoms in a Hubbard-regime optical lattice. Nature **462**, 74 (2009)
5. Bertoldi, A., Lamporesi, G., Cacciapuoti, L., de Angelis, M., Fattori, M., Petelski, T., Peters, A., Prevedelli, M., Stuhler, J., Tino, G.M.: Atom interferometry gravity-gradiometer for the determination of the Newtonian gravitational constant G. Eur. Phys. J. D **40**, 271 (2006)
6. Bloch, I., Hänsch, T.W., Esslinger, T.: Measurement of the spatial coherence of a trapped Bose gas at the phase transition. Nature **403**, 166 (2000)
7. Carruthers, P., Nieto, M.M.: Phase and angle variables in quantum mechanics. Rev. Mod. Phys. **40**, 411 (1968)
8. Castin, Y., Dalibard, J.: Relative phase of two Bose-Einstein condensates. Phys. Rev. A **55**, 4330 (1997)
9. Chen, C.-C., González Escudero, R., Minár, J., Pasquiou, B., Bennetts, S., Schreck, F.: Continuous Bose-Einstein condensation. Nature **606**, 683 (2022)
10. Chikkatur, A.P., Shin, Y., Leanhardt, A.E., Kielpinski, D., Tsikata, E., Gustavson, T.L., Pritchard, D.E., Ketterle, W.: A continuous source of Bose-Einstein condensed atoms. Science **296**, 2193 (2002)
11. Colella, R., Overhauser, A.W., Werner, S.A.: Observation of gravitationally induced quantum interference. Phys. Rev. Lett. **34**, 1472 (1975)
12. Dalvit, D.A.R., Dziarmaga, J., Onofrio, R.: Measurement-induced squeezing of a Bose-Einstein condensate. Phys. Rev. A **65**, 033620 (2002)

References

13. Davisson, C., Germer, L.H.: The scattering of electrons by a single crystal of Nickel. Nature **119**, 558 (1927)
14. Deh, B., Marzok, C., Slama, S., Zimmermann, C., Courteille, P.W.: Bragg spectroscopy and Ramsey interferometry with an ultracold Fermi gas. Appl. Phys. B **97**, 387 (2009)
15. Dirac, P.A.M.: On the Theory of Quantum Mechanics. Proc. Royal Society A **112**, 661 (1926)
16. Feynman, R.P.: Quantum mechanical computers. Opt. News **11**, 11 (1985)
17. Fixler, J.B., Foster, G.T., McGuirk, J.M., Kasevich, M.A.: Atom interferometer measurement of the Newtonian constant of gravity. Science **315**, 74 (2007)
18. Fray, S., Diez, C.A., Hänsch, T.W., Weitz, M.: Atomic interferometer with amplitude gratings of light and its applications to atom based tests of the equivalence principle. Phys. Rev. Lett. **93**, 240404 (2004)
19. Giltner, D.M., McGowan, R.W., Lee, S.A.: Atom interferometer based on Bragg scattering from standing light waves. Phys. Rev. Lett. **75**, 2638 (1995)
20. Glauber, R.J.: Dirac's famous dictum of interference: One photon or two?. Am. J. Phys. **63**, 12 (1995)
21. Gross, C., Zibold, T., Nicklas, E., Estève, J., Oberthaler, M.K.: Nonlinear atom interferometer surpasses classical precision limit. Nature **464**, 1165 (2010)
22. Hagley, E.W., Deng, L., Kozuma, M., Wen, J., Helmerson, K., Rolston, S.L., Phillips, W.D.: A well-collimated quasi-continuous atom laser. Science **283**, 1706 (1999)
23. Helmerson, K., Hutchinson, D., Burnett, K., Phillips, W.D.: Atom lasers. Phys. World **12**(8), 31 (1999)
24. Hosten, O., Engelsen, N.J., Krishnakumar, R., Kasevich, M.A.: Measurement noise 100 times lower than the quantum-projection limit using entangled atoms. Nature **529**, 505 (2016)
25. Inouye, S., Chikkatur, A.P., Stamper-Kurn, D.M., Stenger, J., Pritchard, D.E., Ketterle, W.: Superradiant Rayleigh scattering from a Bose-Einstein condensate. Science **285**, 571 (1999)
26. Javanainen, J., Yoo, S.M.: Quantum phase of a Bose-Einstein condensate with an arbitrary number of atoms. Phys. Rev. Lett. **76**, 161 (1996)
27. Jaklevic, R.C., Lambe, J., Silver, A.H., Mercereau, J.E.: Quantum interference effects in Josephson tunneling. Phys. Rev. Lett. **12**, 159 (1964)
28. Josephson, B.D.: Possible new effects in superconductive tunnelling. Phys. Lett. **1**, 251 (1962)
29. Kozuma, M., Deng, L., Hagley, E.W., Wen, J., Lutwak, R., Helmerson, K., Rolston, S.L., Phillips, W.D.: Coherent splitting of Bose-Einstein condensed atoms with optically induced Bragg diffraction. Phys. Rev. Lett. **82**, 871 (1999)
30. Kozuma, M., Suzuki, Y., Torii, Y., Sugiura, T., Kuga, T.,1 Hagley, E.W., Deng, L.: Phase-coherent amplification of matter waves. Science **286**, 2309 (1999)
31. Lamporesi, G., Bertoldi, A., Cacciapuoti, L., Prevedelli, M., Tino, G.M.: Determination of the Newtonian gravitational constant using atom interferometry. Phys. Rev. Lett. **100**, 050801 (2008)
32. Louradour, F., Reynaud, F., Colombeau, B., Froehly, C.: Interference fringes between two separate lasers. Am. J. Phys. **61**, 242 (1993)
33. Magyar, G., Mandel, L.: Interference fringes produced by superposition of two independent maser light beams. Nature **198**, 255 (1963)
34. Mandel, O., Greiner, M., Widera, A., Rom, T., Hänsch, T.W., Bloch, I.: Controlled collisions for multiparticle entanglement of optically trapped atoms. Nature **425**, 937 (2003)
35. Martin, P.J., Oldaker, B.G., Miklich, A.H., Pritchard, D.E.: Bragg scattering of atoms from a standing light wave. Phys. Rev. Lett. **60**, 515 (1988)
36. Marton, L.: Electron interferometer. Phys. Rev. **85**, 1057 (1952)
37. Michelson, A.A., Morley, E.W.: On the relative motion of the Earth and the luminiferous ether. Am. J. Sci. **34**, 333 (1887)
38. Nelson, K.D., Li, X., Weiss, D.S.: Imaging single atoms in a three-dimensional array. Nat. Phys. **3**, 556 (2007)
39. Orzel, C., Tuchman, A.K., Fenselau, M.L., Yasuda, M., Kasevich, M.A.: Squeezed states in a Bose-Einstein condensate. Science **291**, 2386 (2001)

40. Padley, S., Mas, H., Drougakis, G., Thekkeppatt, P., Bolpasi, V., Vasilakis, G., Poulios, K., von Klitzing, W.: Hypersonic Bose-Einstein condensates in accelerator rings. Nature **570**, 205 (2019)
41. Peters, A., Chung, K.Y., Chu, S.: Measurement of gravitational acceleration by dropping atoms. Nature **400**, 849 (1999)
42. Ramanathan, A., Wright, K.C., Muniz, S.R., Zelan, M., Hill, W.T., III, Lobb, C.J., Helmerson, K., Phillips, W.D., Campbell, G.K.: Superflow in a toroidal Bose-Einstein condensate: An atom circuit with a tunable weak link. Phys. Rev. Lett. **106**, 130401 (2011)
43. Rauch, H., Treimer, W., Bonse, U.: Test of a single crystal neutron interferometer. Phys. Lett. **47**, 369 (1974)
44. Rosi, G., D'Amico, G., Cacciapuoti, L., Sorrentino, F., Prevedelli, M., Zych, M., Brukner, C., Tino, G.M.: Quantum test of the equivalence principle for atoms in coherent superposition of internal energy states. Nature Comm. **8**, 15529 (2017)
45. Shin, Y., Saba, M., Pasquini, T.A., Ketterle, W., Pritchard, D.E., Leanhardt, A.E.: Atom interferometry with Bose-Einstein condensates in a double-well potential. Phys. Rev. Lett. **92**, 050405 (2005)
46. Smerzi, A., Fantoni, S., Giovanazzi, S., Shenoy, S.R.: Quantum coherent atomic tunneling between two trapped Bose-Einstein condensates. Phys. Rev. Lett. **79**, 4950 (1997)
47. Stenger, J., Inouye, S., Chikkatur, A.P., Stamper-Kurn, D.M., Pritchard, D.E., Ketterle, W.: Bragg spectroscopy of a Bose-Einstein condensate. Phys. Rev. Lett. **82**, 4569 (1999)
48. Torrontegui, E., Ibáñez, S., Chen, X., Ruschhaupt, A., Guéry-Odelin, D., Muga, J.G.: Fast atomic transport without vibrational heating. Phys. Rev. A **83**, 013415 (2011)
49. Wright, K.C., Blakestad, R.B., Lobb, C.J., Phillips, W.D., Campbell, G.K.: Driving phase slips in a superfluid atom circuit with a rotating weak link. Phys. Rev. Lett. **110**, 025302 (2013)

Epilogue

Some final considerations are in order. In terms of what we have covered in this book, the first two chapters have been devoted to describe the theoretical and experimental tools to achieve control of external degrees of freedom of atoms to allow their manipulation under conditions in which a quantum description is needed. The initial motivation for this effort was to shed light on a microscopic description of liquid ^{4}He and more in general to have an experimental platform to test mean field theories, as described in the third chapter. Ultracold atom physics has exceeded this initial goal, as described in the fourth chapter for its impact in strongly correlated systems and phase transitions, and in the fifth chapter for the progress induced in metrology and computation. These days we can envision broader implications of this subfield, also on instructional aspects. The possibility to discuss quantum states in a more pristine environment with respect to traditional condensed matter physics, a sort of low-density, low temperature condensed matter physics, certainly could contribute to a better understanding, starting from the fact that the effects of macroscopic wave functions may be literally visualized through densities and velocities of the atoms.

It may be also worth to briefly discuss what we have not covered, partially to contain the size of this book. The subfield of ultracold atoms has growing ramifications to existing fields of research, and we mention at least some which are relatively easy to grasp after studying this book, with further suggested reading. Ultracold systems such as spinor Bose-Einstein condensates, in which several hyperfine components are present, or their fermionic counterpart of unbalanced spin components, are a subject of active investigations with impact on magnetic phases, mixture miscibility, and unconventional superconductivity. Bose-Fermi mixtures, in analogy to ^{3}He-^{4}He mixtures, also fall within this direction. A second subject of research under study is the one of atoms manifesting long-range interatomic interactions, such as gases in which dipole-dipole interactions between atoms prevail over the zero-range pseudopotentials. These dipolar gases can undergo elastic scattering at any temperature, circumventing the bottleneck of fermion cooling in the deep degenerate regime, and allow to study Casimir-Polder forces important in nanotechnology. Finally, we have not described systems which are within reach, both in terms of theoretical and experimental tools, such as Rydberg atoms and ions. Incidentally, both these systems are considered crucial both for metrological purposes and as platforms for efficient quantum computation.

Further Reading

- M. A. Nielsen and I. L. Chuang, Quantum Computation and Quantum Information (1st ed.), (Cambridge University Press, 2000).
- D. Leibfried, R. Blatt, C. Monroe, and D. Wineland, Quantum dynamics of single trapped ions, Rev. Mod. Phys. **75**, 281 (2003).
- K. Blaum, High-accuracy mass spectrometry with stored ions, Phys. Rep. **425**, 1 (2006).
- M. Le Bellac, A Short Introduction to Quantum Information and Quantum Computation (Cambridge Univ. Press, Cambridge, 2006).
- T. Lahaye, C. Menotti, L. Santos, M. Lewenstein, and T. Pfau, The physics of dipolar bosonic quantum gases, Rep. Progr. Phys. **72**, 126401 (2009).
- D. M. Stamper-Kurn and M. Ueda, Spinor Bose gases: Symmetries, magnetism, and quantum dynamics, Rev. Mod. Phys. **85**, 1191 (2013).
- M. Tomza, K. Jachymski, R. Gerritsma, A. Negretti, T. Calarco, Z. Idziaszek, and P. S. Julienne, Cold hybrid ion-atom systems Rev. Mod. Phys. **91**, 035001 (2019).
- X. Wu, X. Liang, Y. Tian, F. Yang, C. Chen, Y.-C. Liu, M. Khoon Tey, and L. You, A concise review of Rydberg atom based quantum computation and quantum simulation, Chin. Phys. B **30**, 020305 (2021).
- X.-F. Shi, Quantum logic and entanglement by neutral Rydberg atoms: methods and fidelity, Quantum Sci. Technol. **7**, 023002 (2022).
- T.F. Gallagher, Rydberg Atoms, in Springer Handbook of Atomic, Molecular, and Optical Physics, G. W. F. Drake (Ed.) (Springer Nature Switzerland, 2023).
- R. Aguado, R. Citro, M. Lewenstein, M. Stern, New trends and platforms for quantum technologies (Springer, 2024).

Second Quantization and Many-Body Systems   A

We describe here the second quantization approach to nonrelativistic many-body systems. One could argue that the two concepts are at variance, as second quantization has been introduced to describe relativistic systems in which particles can be created and destroyed at will. However, at least three reasons make this discussion logically acceptable. Certainly, in many-body systems, there are quantum states which can be created and destroyed as well, both collective excitations and quasiparticles belong to this realm. The formal application of second quantization in these degrees of freedom is therefore warranted. Second, the description of a system in quantum mechanics is in terms of states, not necessarily particles in themselves. States are operatively defined, for the most complete description, by the maximal number of eigenvalues of compatible observables. A state is different from another state if it has at least a different eigenvalue. In this sense, a free electron described with a momentum eigenvalue p and a third component of the spin σ_z is different from a free electron with momentum eigenvalue p' and a third component of the spin σ_z'. If the first electron undergoes a scattering process or a spin-flip process ending up having momentum p' and third component of the spin σ_z', we can describe the process as having destroyed an electron in state $|p; \sigma_z\rangle$ and having created an electron in state $|p'; \sigma_z'\rangle$. In this framework, even if the electron as a particle is neither destroyed not created, the states associated to it are! Finally, second quantization allows to write in an elegant way many-body quantities, such as the state of a system, and the related Hamiltonian operator, in a way that is both aesthatically pleasant and computationally efficient, incorporating at the same time the principle of indistinguishability of identical particles.

The second quantization approach starts from describing a many-body quantum system in terms of occupation numbers of particles in defined states labeled by eigenvalues of observables. This state is defined in a generalization of the Hilbert space which allows for an arbitrary number of particles, called Fock space. It is clear from the outset that this choice solves all the issues of indistinguishability between the particles, provided that proper care is taken in the case of fermions limiting the occupation number to unity for each state. Then, symmetrization or

© The Author(s), under exclusive license to Springer Nature Switzerland AG 2024
R. Onofrio, L. Salasnich, *Physics and Technology of Ultracold Atomic Gases*,
Lecture Notes in Physics 1034, https://doi.org/10.1007/978-3-031-76004-4

antisymmetrization protocols as in first quantization are completely bypassed. A generic state of this Fock space is given by

$$|\ldots n_\alpha \ldots n_\beta \ldots n_\gamma \ldots\rangle, \quad (A.1)$$

indicating the presence of n_α particles in the single-particle state identified by $|\alpha\rangle$, n_β particles in the single-particle state $|\beta\rangle$, n_γ particles in the single-particle state $|\gamma\rangle$, etc. A special role among all Fock states is played by the vacuum state, with zero occupation number in each mode:

$$|0\rangle = |\ldots 0 \ldots 0 \ldots 0 \ldots\rangle. \quad (A.2)$$

The dynamics in the Fock space will be determined by changes in the occupation numbers, with which particles can be moved from one state to another, created and destroyed, keeping in mind that it is impossible to destroy particles in the vacuum state. This requires the introduction of annihilation and creation operators whose form for removing or adding particles in state α will be chosen as $\hat{c}_\alpha$ and $\hat{c}_\alpha^+$. The action of these operators on a ket state is as follows:

$$\hat{c}_\alpha |\ldots n_\alpha \ldots\rangle = \sqrt{n_\alpha} |\ldots n_\alpha - 1 \ldots\rangle, \quad (A.3)$$

$$\hat{c}_\alpha^+ |\ldots n_\alpha \ldots\rangle = \sqrt{n_\alpha + 1} |\ldots n_\alpha + 1 \ldots\rangle, \quad (A.4)$$

and their action on the dual (bra) state is

$$\langle\ldots n_\alpha \ldots| \hat{c}_\alpha^+ = \sqrt{n_\alpha} \langle\ldots n_\alpha - 1 \ldots|, \quad (A.5)$$

$$\langle\ldots n_\alpha \ldots| \hat{c}_\alpha = \sqrt{n_\alpha + 1} \langle\ldots n_\alpha + 1 \ldots|, \quad (A.6)$$

with also $\hat{c}_\alpha |0\rangle = 0$, $\langle 0| \hat{c}_\alpha^+ = 0$. Since all these states form an orthonormal basis, by taking the inner products of bras and kets, we can write

$$\langle\ldots n_\alpha \ldots| \hat{c}_\alpha^+ \hat{c}_\alpha |\ldots n_\alpha \ldots\rangle = n_\alpha \langle\ldots n_\alpha - 1 \ldots|\ldots n_\alpha - 1 \ldots\rangle = n_\alpha, \quad (A.7)$$

$$\langle\ldots n_\alpha \ldots| \hat{c}_\alpha \hat{c}_\alpha^+ |\ldots n_\alpha \ldots\rangle = (n_\alpha + 1)\langle\ldots n_\alpha + 1 \ldots|\ldots n_\alpha + 1 \ldots\rangle$$
$$= n_\alpha + 1. \quad (A.8)$$

This implies two important relationships. From Eq. (A.7), we can identify a number operator for occupation state α as

$$\hat{n}_\alpha = \hat{c}_\alpha^+ \hat{c}_\alpha, \quad (A.9)$$

A Second Quantization and Many-Body Systems

with eigenvalue n_α. By subtracting Eq. (A.7) from Eq. (A.8), we obtain

$$\hat{c}_\alpha \hat{c}_\alpha^+ - \hat{c}_\alpha^+ \hat{c}_\alpha = [\hat{c}_\alpha, \hat{c}_\alpha^+] = 1. \qquad (A.10)$$

Also, annihilation and creation operators acting on different levels α, β satisfy

$$[\hat{c}_\alpha, \hat{c}_\beta] = [\hat{c}_\alpha^+, \hat{c}_\beta^+] = 0, \quad [\hat{c}_\alpha, \hat{c}_\beta^+] = \delta_{\alpha\beta}. \qquad (A.11)$$

This certainly holds for bosons, as in this case we can pile up indefinite number of particles in any state. In the case of fermions, we need to take into account the fact that no more than one particle can occupy a single state. Therefore, the only possibilities for the annihilation and creation operators are the following:

$$\hat{c}_\alpha | \ldots 1_\alpha \ldots \rangle = | \ldots 0_\alpha \ldots \rangle, \qquad (A.12)$$

$$\hat{c}_\alpha^+ | \ldots 0_\alpha \ldots \rangle = | \ldots 1_\alpha \ldots \rangle, \qquad (A.13)$$

and their action on the dual (bra) state is

$$\langle \ldots 1_\alpha \ldots | \hat{c}_\alpha^+ = \langle \ldots 0_\alpha \ldots |, \qquad (A.14)$$

$$\langle \ldots 0_\alpha \ldots | \hat{c}_\alpha = \langle \ldots 1_\alpha \ldots |. \qquad (A.15)$$

Let us consider the following inner products:

$$\langle \ldots 0_\alpha \ldots | \hat{c}_\alpha \hat{c}_\alpha^+ | \ldots 0_\alpha \ldots \rangle = \langle \ldots 1_\alpha \ldots | \ldots 1_\alpha \ldots \rangle = 1, \qquad (A.16)$$

$$\langle \ldots 1_\alpha \ldots | \hat{c}_\alpha^+ \hat{c}_\alpha | \ldots 1_\alpha \ldots \rangle = \langle \ldots 0_\alpha \ldots | \ldots 0_\alpha \ldots \rangle = 1, \qquad (A.17)$$

$$\langle \ldots 0_\alpha \ldots | \hat{c}_\alpha^+ \hat{c}_\alpha | \ldots 0_\alpha \ldots \rangle = 0, \qquad (A.18)$$

$$\langle \ldots 1_\alpha \ldots | \hat{c}_\alpha \hat{c}_\alpha^+ | \ldots 1_\alpha \ldots \rangle = 0. \qquad (A.19)$$

The relationships in Eqs. (A.16)–(A.17)–(A.18) are also common to the bosonic case. Equation (A.19) instead contains the impossibility of having more than one fermion per state. If we were to impose here the commutation relation of Eq. (A.10), we would have

$$0 = \langle \ldots 1_\alpha \ldots | \hat{c}_\alpha \hat{c}_\alpha^+ | \ldots 1_\alpha \ldots \rangle = \langle \ldots 1_\alpha \ldots | \hat{c}_\alpha^+ \hat{c}_\alpha | \ldots 1_\alpha \ldots \rangle + 1$$

$$= \langle \ldots 0_\alpha \ldots | \ldots 0_\alpha \ldots \rangle + 1 = 2, \qquad (A.20)$$

which is a manifest contradiction. This is solved if we instead impose an anticommutation relationship changing the sign of the $c_\alpha^+ c_\alpha$ term:

$$\hat{c}_\alpha \hat{c}_\alpha^+ + \hat{c}_\alpha^+ \hat{c}_\alpha = \{\hat{c}_\alpha, \hat{c}_\alpha^+\} = 1. \qquad (A.21)$$

Moreover, if annihilation or creation operators are applied twice, the outcome should be zero:

$$(\hat{c}_\alpha)^2 = 0, \quad (\hat{c}_\alpha^+)^2 = 0. \tag{A.22}$$

In this way, we can also verify that the eigenvalues of the number operator, still defined as in Eq. (A.9), are only 0 and 1, since we have

$$\hat{n}_\alpha^2 = \left(\hat{c}_\alpha^+ \hat{c}_\alpha\right)^2 = \hat{c}_\alpha^+ \hat{c}_\alpha \hat{c}_\alpha^+ \hat{c}_\alpha = \hat{c}_\alpha^+ \left(1 - \hat{c}_\alpha^+ \hat{c}_\alpha\right) \hat{c}_\alpha = \hat{c}_\alpha^+ \hat{c}_\alpha = \hat{n}_\alpha. \tag{A.23}$$

Also, for a two-particle state, we have

$$|1_\alpha 1_\beta\rangle = \hat{c}_\alpha^+ \hat{c}_\beta^+ |0\rangle = -\hat{c}_\beta^+ \hat{c}_\alpha^+ |0\rangle = -|1_\beta 1_\alpha\rangle, \tag{A.24}$$

i.e., the state is antisymmetric under interchange of particle labels, showing that the formalism has built in the antisymmetric request under particle exchange.

So far, the description is general and as such is applicable to both relativistic or nonrelativistic systems. In the former, the dispersion relationship for a free particle of mass m is $E^2 = p^2 c^2 + m^2 c^4$ while in the nonrelativistic limit is approximated as $E = p^2/(2m)$. Therefore, in the first case, we expect negative energy solutions associated to the presence of antiparticles, while these are absent in the nonrelativistic case. Also, in a relativistic setting, at least in the case of scattering problems, the natural observables are momenta of the particles, while in the nonrelativistic setting in general, the emphasis is on the spatial position of the particles.

Indeed, the nonrelativistic description of the single-particle state at a given time may be expressed in terms of a function of position and, possibly, the eigenvalues of other compatible observables, via the wave function:

$$\phi_\alpha(\mathbf{r}) = \langle \mathbf{r} | \alpha \rangle. \tag{A.25}$$

This single particle wave function is the solution of the nonrelativistic time-independent Schrödinger equation:

$$\hat{H} \phi_\alpha(\mathbf{r}) = \epsilon_\alpha \phi_\alpha(\mathbf{r}), \tag{A.26}$$

where we have introduced the Hamiltonian operator

$$\hat{H} = -\frac{\hbar^2}{2m} \nabla^2 + U(\mathbf{r}), \tag{A.27}$$

with energy eigenvalues ϵ_α and $U(\mathbf{r})$ is a time-independent potential energy due to an external conservative field. The creation and destruction of particles characterized by the state $|\alpha\rangle$ in a specific location of space $\mathbf{r}$ will be achieved by combining the creation and destruction operators with the functions $\phi_\alpha(\mathbf{r})$. We then define the field operator or quantized field as

$$\hat{\psi}(\mathbf{r}) = \sum_\alpha \hat{c}_\alpha \, \phi_\alpha(\mathbf{r}), \tag{A.28}$$

and its Hermitian conjugate will be then

$$\hat{\psi}^+(\mathbf{r}) = \sum_\alpha \hat{c}_\alpha^+ \, \phi_\alpha^*(\mathbf{r}). \tag{A.29}$$

These operators have a hybrid task. They act as operators in the Fock space, annihilating and creating particles in state $|\alpha\rangle$, respectively, but at the same time, the wave function present in it declares the probability amplitude with which such processes are taking place in a spatial point $\mathbf{r}$ (or, in general, in any space of observables). Therefore, on top of the operatorial structure of nonrelativistic quantum mechanics with operators acting on wave functions, we now have a new layer of quantization, with local, space-dependent, operators acting in the Fock space. This is the origin of the term second quantization.

Also, notice that if $\hat{c}_\alpha$ and $\hat{c}_\alpha^+$ become numbers, these can be interpreted as the probability amplitudes of the corresponding eigenstate $\phi_\alpha(\mathbf{r})$, i.e., $c_\alpha = \langle \alpha | \psi \rangle$. Therefore, the relationship between second quantization and first quantization is obtained with the following correspondence:

First quantization $\Longleftrightarrow$ Second quantization

$$c_\alpha \Longleftrightarrow \hat{c}_\alpha \tag{A.30}$$

$$\psi(\mathbf{r}) \Longleftrightarrow \hat{\psi}(\mathbf{r}) \tag{A.31}$$

$$c_\alpha^* \Longleftrightarrow \hat{c}_\alpha^+ \tag{A.32}$$

$$\psi^*(\mathbf{r}) \Longleftrightarrow \hat{\psi}^+(\mathbf{r}) \tag{A.33}$$

This correspondence is useful in both ways. Given the quantities in first quantization, one can promote them to become second quantization quantities or, alternatively, can demote second quantization quantities to become their corresponding ones in first quantization, a sort of "classical" limit from the second quantization standpoint.

An important property of the field operator $\hat{\psi}^+(\mathbf{r})$ is the following:

$$\hat{\psi}^+(\mathbf{r})|0\rangle = |\mathbf{r}\rangle, \tag{A.34}$$

that is, the operator $\hat{\psi}^+(\mathbf{r})$ creates a particle in the state $|\mathbf{r}\rangle$ from the vacuum state $|0\rangle$.

In fact, considering that $|\alpha\rangle$ states fulfil completeness, we have

$$\hat{\psi}^+(\mathbf{r})|0\rangle = \sum_\alpha \hat{c}_\alpha^+ \phi_\alpha^*(\mathbf{r})|0\rangle = \sum_\alpha \hat{c}_\alpha^+ \langle\mathbf{r}|\alpha\rangle^*|0\rangle = \sum_\alpha \langle\alpha|\mathbf{r}\rangle|\alpha\rangle$$

$$= \sum_\alpha |\alpha\rangle\langle\alpha|\mathbf{r}\rangle = |\mathbf{r}\rangle . \tag{A.35}$$

Since $1 = \langle\mathbf{r}|\mathbf{r}\rangle = \langle\mathbf{r}|\hat{\psi}^+(\mathbf{r})|0\rangle = \langle 0|\hat{\psi}^+(\mathbf{r})|\mathbf{r}\rangle^* = \langle 0|\hat{\psi}(\mathbf{r})|\mathbf{r}\rangle = \langle 0|0\rangle = 1$, we also have the reverse process of Eq. (A.34), the destruction of a particle in the state $|\mathbf{r}\rangle$

$$\hat{\psi}(\mathbf{r})|\mathbf{r}\rangle = |0\rangle . \tag{A.36}$$

Based on Eqs. (A.10)–(A.11) for bosons, the corresponding field operators satisfy

$$[\hat{\psi}(\mathbf{r}), \hat{\psi}^+(\mathbf{r}')] = \delta(\mathbf{r} - \mathbf{r}') , \tag{A.37}$$

while using Eqs. (A.21)–(A.22) for fermions, one gets

$$\{\hat{\psi}(\mathbf{r}), \hat{\psi}^+(\mathbf{r}')\} = \delta(\mathbf{r} - \mathbf{r}') . \tag{A.38}$$

Indeed, by using the expansion of the field operators, we have

$$[\hat{\psi}(\mathbf{r}), \hat{\psi}^+(\mathbf{r}')]$$
$$= \sum_{\alpha,\beta} \phi_\alpha(\mathbf{r}) \phi_\beta^*(\mathbf{r}') [c_\alpha, c_\beta^+] = \sum_{\alpha,\beta} \phi_\alpha(\mathbf{r}) \phi_\beta^*(\mathbf{r}') \delta_{\alpha,\beta} = \sum_\alpha \phi_\alpha(\mathbf{r}) \phi_\alpha^*(\mathbf{r}')$$
$$= \sum_\alpha \langle\mathbf{r}|\alpha\rangle\langle\alpha|\mathbf{r}'\rangle = \langle\mathbf{r}|\sum_\alpha |\alpha\rangle\langle\alpha|\mathbf{r}'\rangle = \langle\mathbf{r}|\mathbf{r}'\rangle = \delta(\mathbf{r} - \mathbf{r}') . \tag{A.39}$$

The proof for fermions is similar, just using anticommutators $\{c_\alpha, c_\beta^+\}$ instead of the commutators $[c_\alpha, c_\beta^+]$ in Eq. (A.39).

Now, we focus our attention on two important local quantities in second quantization. We look for the analogous of the probability density of first quantization, $\rho(\mathbf{r}) = \psi^*(\mathbf{r})\psi(\mathbf{r})$, and using the correspondence in Eqs. (A.31)–(A.33), we introduce a density operator as

$$\hat{\rho}(\mathbf{r}) = \hat{\psi}^+(\mathbf{r})\hat{\psi}(\mathbf{r}) = \sum_{\alpha,\beta} \hat{c}_\alpha^+ \hat{c}_\beta \, \phi_\alpha^*(\mathbf{r}) \phi_\beta(\mathbf{r}) , \tag{A.40}$$

A Second Quantization and Many-Body Systems

with the total number operator written in terms of $\hat{\rho}(\mathbf{r})$ as

$$\hat{N} = \int d^3\mathbf{r}\, \hat{\rho}(\mathbf{r}). \tag{A.41}$$

The density operator $\hat{\rho}(\mathbf{r})$, using commutation or anticommutation relations of the field operators, may be shown to be related to the probability density as

$$\hat{\rho}(\mathbf{r})\,|\mathbf{r}_1\mathbf{r}_2\ldots\mathbf{r}_N\rangle = \sum_{i=1}^{N} \delta(\mathbf{r}-\mathbf{r}_i)\,|\mathbf{r}_1\mathbf{r}_2\ldots\mathbf{r}_N\rangle, \tag{A.42}$$

where we can identify $\rho(\mathbf{r}) = \sum_{i=1}^{N} \delta(\mathbf{r}-\mathbf{r}_i)$ as the one-body density function for a system of N particles located in $\mathbf{r}_i$.

Similarly, the Hamiltonian operator in Eq. (A.27) allows to evaluate the average energy for a particle in a given state as

$$E = \langle\psi|\hat{H}|\psi\rangle = \int d^3\mathbf{r}\, \psi^*(\mathbf{r})\left[-\frac{\hbar^2}{2m}\nabla^2 + U(\mathbf{r})\right]\psi(\mathbf{r}). \tag{A.43}$$

This is replaced in second quantization by an Hamiltonian operator acting on the Fock states, defined as

$$\hat{H} = \int d^3\mathbf{r}\, \hat{\psi}^+(\mathbf{r},t)\left[-\frac{\hbar^2}{2m}\nabla^2 + U(\mathbf{r})\right]\hat{\psi}(\mathbf{r},t). \tag{A.44}$$

By using Eqs. (A.28)–(A.29) and the orthonormality of the basis of energy eigenstates $\{\phi_\alpha\}$, Eq. (A.44) becomes

$$\hat{H} = \sum_\alpha \epsilon_\alpha \hat{N}_\alpha = \sum_\alpha \epsilon_\alpha\, \hat{c}_\alpha^+ \hat{c}_\alpha. \tag{A.45}$$

The extension to the case of interacting many-body systems is immediate. In first quantization, the Hamiltonian operator of N interacting and identical particles in the external potential $U(\mathbf{r})$ is given by

$$\hat{H}^{(N)} = \sum_{i=1}^{N}\left[-\frac{\hbar^2}{2m}\nabla_i^2 + U(\mathbf{r}_i)\right] + \frac{1}{2}\sum_{\substack{i,j=1\\i\neq j}} V(\mathbf{r}_i-\mathbf{r}_j) = \sum_{i=1}^{N}\hat{h}_i + \frac{1}{2}\sum_{\substack{i,j=1\\i\neq j}}^{N} V_{ij}. \tag{A.46}$$

where $V(\mathbf{r}-\mathbf{r}')$ is the interparticle potential.

The second quantized Hamiltonian operator will be similar to the one in Eq. (A.44), being simply augmented by the interparticle potential in which two-body operators will appear

$$\hat{H} = \int d^3\mathbf{r}\, \hat{\psi}^+(\mathbf{r}) \left[-\frac{\hbar^2}{2m}\nabla^2 + U(\mathbf{r}) \right] \hat{\psi}(\mathbf{r})$$

$$+ \frac{1}{2} \int d^3\mathbf{r}\, d^3\mathbf{r}'\, \hat{\psi}^+(\mathbf{r})\, \hat{\psi}^+(\mathbf{r}')\, V(\mathbf{r}-\mathbf{r}')\, \hat{\psi}(\mathbf{r}')\, \hat{\psi}(\mathbf{r}) . \quad \text{(A.47)}$$

This can be rewritten in terms of creation and annihilation operators, using the same base of eigenstates for the corresponding noninteracting single-particle, as

$$\hat{H} = \sum_\alpha \epsilon_\alpha\, \hat{c}^+_\alpha \hat{c}_\alpha + \sum_{\alpha\beta\gamma\delta} V_{\alpha\beta\gamma\delta}\, \hat{c}^+_\alpha \hat{c}^+_\beta \hat{c}_\delta \hat{c}_\gamma , \quad \text{(A.48)}$$

where

$$V_{\alpha\beta\delta\gamma} = \int d^3\mathbf{r}\, d^3\mathbf{r}'\, \phi^*_\alpha(\mathbf{r})\, \phi^*_\beta(\mathbf{r}')\, V(\mathbf{r}-\mathbf{r}')\, \phi_\delta(\mathbf{r}')\, \phi_\gamma(\mathbf{r}) . \quad \text{(A.49)}$$

Even in this case, in analogy to Eq. (A.42), there is a connection between second quantization Hamiltonian $\hat{H}$ and the first quantization Hamiltonian $\hat{H}^{(N)}$, given by the formula:

$$\hat{H}|\mathbf{r}_1\mathbf{r}_2\ldots\mathbf{r}_N\rangle = \hat{H}^{(N)}|\mathbf{r}_1\mathbf{r}_2\ldots\mathbf{r}_N\rangle . \quad \text{(A.50)}$$

In fact, one finds that

$$\hat{\psi}^+(\mathbf{r})\,\hat{h}(\mathbf{r})\,\hat{\psi}(\mathbf{r})\,|\mathbf{r}_1\mathbf{r}_2\ldots\mathbf{r}_N\rangle = \sum_{i=1}^N \hat{h}(\mathbf{r}_i)\delta(\mathbf{r}-\mathbf{r}_i)\,|\mathbf{r}_1\mathbf{r}_2\ldots\mathbf{r}_N\rangle , \quad \text{(A.51)}$$

and also

$$\hat{\psi}^+(\mathbf{r})\,\hat{\psi}^+(\mathbf{r}')\,V(\mathbf{r},\mathbf{r}')\,\hat{\psi}(\mathbf{r}')\,\hat{\psi}(\mathbf{r})\,|\mathbf{r}_1\mathbf{r}_2\ldots\mathbf{r}_N\rangle$$

$$= \sum_{\substack{i,j=1 \\ i\neq j}}^N V(\mathbf{r}_i,\mathbf{r}_i)\delta(\mathbf{r}-\mathbf{r}_i)\,\delta(\mathbf{r}'-\mathbf{r}_j)\,|\mathbf{r}_1\mathbf{r}_2\ldots\mathbf{r}_N\rangle . \quad \text{(A.52)}$$

From these two expressions, Eq. (A.50) follows after integration over spatial coordinates.

A Second Quantization and Many-Body Systems

Finally, we discuss the time evolution. The time-dependent equation of motion for the field operator $\hat{\psi}(\mathbf{r}, t)$ is easily obtained from the Hamiltonian (A.47) by using the Heisenberg equation:

$$i\hbar \frac{\partial}{\partial t} \hat{\psi} = [\hat{\psi}, \hat{H}], \qquad (A.53)$$

which gives, in the case of the interacting many-body system described by Eq. (A.47)

$$i\hbar \frac{\partial}{\partial t} \hat{\psi} = \left[-\frac{\hbar^2}{2m} \nabla^2 + U(\mathbf{r}) \right] \hat{\psi}(\mathbf{r}) + \int d^3\mathbf{r}' \, \hat{\psi}^+(\mathbf{r}') \, V(\mathbf{r} - \mathbf{r}') \, \hat{\psi}(\mathbf{r}') \, \hat{\psi}(\mathbf{r}) . \qquad (A.54)$$

This equation for the field operator $\hat{\psi}(\mathbf{r}, t)$ is the second-quantized version of the Hartree-like time-dependent nonlinear Schrödinger equation for the wave function $\psi(\mathbf{r}, t)$. Note that the same equation is obtained in terms of variation of $\hat{H}$ with respect to $\delta \hat{\psi}^+$

$$i\hbar \frac{\partial}{\partial t} \hat{\psi} = \frac{\delta \hat{H}}{\delta \hat{\psi}^+}, \qquad (A.55)$$

to determine the Heisenberg equation of motion, where

$$\frac{\delta \hat{H}}{\delta \hat{\psi}^+} = \left(\frac{\delta H}{\delta \psi^*} \right)_{\psi = \hat{\psi}, \psi^* = \hat{\psi}^+}. \qquad (A.56)$$

The equation of motion for the time-dependent field operator $\hat{\psi}^+(\mathbf{r}, t)$ is obtained in the same way.

Similarly, the time-dependent equation of motion of the field operator $\hat{c}_\alpha(t)$ is obtained from the Hamiltonian (A.49) by using the Heisenberg equation:

$$i\hbar \frac{d}{dt} \hat{c}_\alpha = [\hat{c}_\alpha, \hat{H}], \qquad (A.57)$$

which gives

$$i\hbar \frac{d}{dt} \hat{c}_\alpha = \hat{c}_\alpha + \sum_{\beta\gamma\delta} V_{\alpha\beta\gamma\delta} \, \hat{c}_\beta^+ \hat{c}_\delta \hat{c}_\gamma . \qquad (A.58)$$

Again, one can obtain the same equation by using the simpler formula:

$$i\hbar \frac{d}{dt}\hat{c}_\alpha = \frac{\partial \hat{H}}{\partial \hat{c}_\alpha^+},\qquad\text{(A.59)}$$

where

$$\frac{\partial \hat{H}}{\partial \hat{a}_\alpha^+} = \left(\frac{\partial H}{\partial a_\alpha^*}\right)_{a_\alpha=\hat{a}_\alpha,\, a_\alpha^*=\hat{a}_\alpha^+}.\qquad\text{(A.60)}$$

Coherent States and Gross-Pitaevskii Equation B

In this Appendix, we derive the Gross-Pitaevskii equation Eq. (1.129) from the Heisenberg equation of motion for the bosonic field operator $\hat{\psi}(\mathbf{r}, t)$ with a local interatomic interaction of strength g:

$$i\hbar \frac{\partial}{\partial t} \hat{\psi}(\mathbf{r}, t) = \left[-\frac{\hbar^2}{2m} \nabla^2 + U(\mathbf{r}) \right] \hat{\psi}(\mathbf{r}, t) + g\, \hat{\psi}^+(\mathbf{r}, t) \hat{\psi}(\mathbf{r}, t) \hat{\psi}(\mathbf{r}, t). \quad (B.1)$$

Let us consider the state $|c_0\rangle$ defined as the eigenstate of the annihilation operator $\hat{c}_0$. Since the annihilation operator is not Hermitian, it will have in general a complex eigenvalue $c_0 = |c_0| e^{i\theta_0}$, such that $\hat{c}_0 |c_0\rangle = c_0 |c_0\rangle$. This state is called "coherent" (see also discussion in Sect. 5.1) and can be expanded in terms of number states $|N, 0, 0, \dots\rangle$:

$$|c_0\rangle = \sum_{N=0}^{\infty} \langle \dots, 0, 0, N | c_0 \rangle | N, 0, 0, \dots \rangle = e^{-|c_0|^2/2} \sum_{N=0}^{\infty} \frac{c_0^N}{\sqrt{N!}} |N, 0, 0, \dots\rangle. \quad (B.2)$$

Evidently, in this coherent state, the number of bosons is not determinate, but using Eq. (A.9), we can calculate the average number of bosons:

$$\bar{N} = \langle c_0 | \hat{N} | c_0 \rangle = |c_0|^2 . \quad (B.3)$$

Coherent states are useful to handle the first quantization wave functions which are the the closest counterparts of the second quantization field operators. Indeed, let us consider again

$$\hat{\psi}(\mathbf{r}, t) = \sum_\alpha \hat{c}_\alpha\, \phi_\alpha(\mathbf{r}, t) , \quad (B.4)$$

and suppose that the generic state is written in terms of coherent states as $|\psi\rangle = \prod_\alpha |c_\alpha\rangle$. Then, we have

$$\hat{\psi}(\mathbf{r})|\psi\rangle = \sum_\alpha \hat{c}_\alpha \, \phi_\alpha(\mathbf{r}, t) \prod_\alpha |c_\alpha\rangle = \sum_\alpha \phi_\alpha(\mathbf{r}, t) \prod_\alpha \hat{c}_\alpha |c_\alpha\rangle$$

$$= \sum_\alpha c_\alpha \phi_\alpha(\mathbf{r}, t) \prod_\alpha |c_\alpha\rangle = \psi(\mathbf{r})|\psi\rangle \,, \tag{B.5}$$

where $\psi(\mathbf{r}) = \sum_\alpha c_\alpha \phi_\alpha(\mathbf{r})$ is the wave function. In synthesis, we have an operatorial equation for the quantized field having as eigenvalue the corresponding first-quantized wave function:

$$\hat{\psi}(\mathbf{r})|\psi\rangle = \psi(\mathbf{r})|\psi\rangle \,. \tag{B.6}$$

Let us now consider Eq. (B.1) calculating its expectation value over a coherent state $|c_0\rangle$, using the expectation value $\langle c_0|\hat{\psi}(\mathbf{r})|c_0\rangle = \bar{N}^{1/2} e^{i\theta_0} \phi_0(\mathbf{r})$. The three terms in Eq. (B.1) are then written as

$$\langle c_0| i\hbar \frac{\partial}{\partial t} \hat{\psi}(\mathbf{r}, t)|c_0\rangle = i\hbar \frac{\partial}{\partial t} \langle c_0|\hat{\psi}(\mathbf{r}, t)|c_0\rangle = c_0 \, i\hbar \frac{\partial}{\partial t} \phi_0(\mathbf{r}, t) \,, \tag{B.7}$$

$$\langle c_0| \left[-\frac{\hbar^2}{2m} \nabla^2 + U(\mathbf{r}) \right] \hat{\psi}(\mathbf{r}, t)|c_0\rangle = \left[-\frac{\hbar^2}{2m} \nabla^2 + U(\mathbf{r}) \right] \langle c_0|\hat{\psi}(\mathbf{r}, t)|c_0\rangle$$

$$= c_0 \left[-\frac{\hbar^2}{2m} \nabla^2 + U(\mathbf{r}) \right] \phi_0(\mathbf{r}, t) \,, \tag{B.8}$$

$$\langle c_0| \hat{\psi}^+(\mathbf{r}, t) \hat{\psi}(\mathbf{r}, t) \hat{\psi}(\mathbf{r}, t)|c_0\rangle$$

$$= \langle c_0| \sum_\alpha \hat{c}_\alpha^+ \phi_\alpha^*(\mathbf{r}, t) \sum_\beta \hat{c}_\beta \, \phi_\beta(\mathbf{r}, t) \sum_\gamma \hat{c}_\gamma \, \phi_\gamma(\mathbf{r}, t) |c_0\rangle$$

$$= |c_0|^2 c_0 |\phi_0(\mathbf{r}, t)|^2 \phi_0(\mathbf{r}, t) \,. \tag{B.9}$$

By equating Eq. (B.7) to the sum of Eqs. (B.8) and (B.9), we conclude that

$$i\hbar \frac{\partial}{\partial t} \phi_0(\mathbf{r}, t) = \left[-\frac{\hbar^2}{2m} \nabla^2 + U(\mathbf{r}) \right] \phi_0(\mathbf{r}, t) + |c_0|^2 |\phi_0(\mathbf{r}, t)|^2 \phi_0(\mathbf{r}, t) \tag{B.10}$$

is the Gross-Pitaevskii equation for $\phi_0(\mathbf{r}, t)$, related to Eq. (1.129) via $\psi(\mathbf{r}, \mathbf{t}) = \bar{N}^{1/2} \phi_0(\mathbf{r}, t)$ and Eq. (B.3).

It is worth to stress that by using the coherent states one can derive the Gross-Pitaevskii equation from the Heisenberg equation (B.1) also in a slightly different way. In particular, instead of using the single-mode coherent state $|c_0\rangle$, one can use the multimode coherent state $|\psi\rangle = \prod_\alpha |c_\alpha\rangle$, satisfying Eq. (B.6). More explicitly, taking into account that $|\psi\rangle$ is normalized, we have

$$\langle\psi|i\hbar\frac{\partial}{\partial t}\hat{\psi}(\mathbf{r},t)|\psi\rangle = i\hbar\frac{\partial}{\partial t}\psi(\mathbf{r},t)\,, \tag{B.11}$$

$$\langle\psi|\left[-\frac{\hbar^2}{2m}\nabla^2 + U(\mathbf{r})\right]\hat{\psi}(\mathbf{r},t)|\psi\rangle = \left[-\frac{\hbar^2}{2m}\nabla^2 + U(\mathbf{r})\right]\psi(\mathbf{r},t)\,, \tag{B.12}$$

$$\langle\psi|g\,\hat{\psi}^+(\mathbf{r},t)\hat{\psi}(\mathbf{r},t)\hat{\psi}(\mathbf{r},t)|\psi\rangle = g\,|\psi(\mathbf{r},t)|^2\psi(\mathbf{r},t)\,, \tag{B.13}$$

also allowing to obtain Eq. (1.129). It is in this sense that the Gross-Pitaevskii equation is sometime considered "classical."

Scattering Theory

In a quantum approach to scattering, one usually assumes that an incoming particle, described by a plane wave along the z-axis, scatters with a target, modelled by an external spherically symmetric potential $V(\mathbf{r})$. The same result is obtained by considering the scattering between two particles of mass m_1, m_2 in the center of mass frame. In this case, the effective particle has the reduced mass $m_r = m_1 m_2/(m_1+m_2)$, and $V(\mathbf{r})$ represents the interparticle potential with $\mathbf{r}$ the relative position vector. In the following, we will consider the collision between two identical particles which is described by the stationary Schrödinger equation:

$$\left(-\frac{\hbar^2}{2m_r}\nabla^2 + V(\mathbf{r})\right)\psi(\mathbf{r}) = E_k\,\psi(\mathbf{r}), \tag{C.1}$$

with $E_k = \hbar^2 k^2/(2m_r)$ the energy of the effective particle characterized by the initial wave vector $\mathbf{k} = (0,0,k)$. An asymptotic solution of Eq. (C.1) reads

$$\psi(\mathbf{r}) \simeq e^{ikz} + f(k,\theta)\frac{e^{ikr}}{r}, \tag{C.2}$$

a superposition between the initial plane wave e^{ikz} along the z axis and the final spherical wave e^{ikr}/r, with $f(k,\theta)$ is called scattering amplitude. The latter determines the angular differential cross section is given by

$$\frac{d\sigma}{d\Omega} = |f(k,\theta)|^2. \tag{C.3}$$

The scattering amplitude $f(k,\theta)$ is often written as

$$f(k,\theta) = \sum_{l=0}^{\infty}(2l+1)f_l(k)\,P_l(\cos(\theta)), \tag{C.4}$$

where $P_l(x)$ are the Legendre polynomials and $f_l(k)$ are the partial-wave scattering amplitudes. Moreover, usually, one introduces the phase shift $\delta_l(k)$ such that

$$f_l(k) = \frac{e^{i\delta_l(k)}}{k} \sin(\delta_l(k)) . \qquad (C.5)$$

The corresponding integrated cross section $\sigma(k)$ reads

$$\sigma(k) = \int \frac{d\sigma}{d\Omega} d\Omega = \int_0^{2\pi} d\phi \int_0^\pi d\theta \, \sin(\theta) \, |f(k,\theta)|^2$$

$$= 4\pi \sum_{l=0}^\infty (2l+1)|f_l(k)|^2 = \frac{4\pi}{k^2} \sum_{l=0}^\infty (2l+1) \sin^2(\delta_l(k)) . \qquad (C.6)$$

At low momenta, one can consider only the s-wave scattering ($l = 0$) and the differential cross section does not depend on the angle θ. In this case, Eq. (C.6) becomes

$$\sigma(k) = 4\pi |f_0(k)|^2 = \frac{4\pi}{k^2} \sin^2(\delta_0(k)) , \qquad (C.7)$$

where $\delta_0(k)$ is the s-wave phase shift of the cross section. The s-wave scattering length a_s, which characterizes the effective size of the target, is then defined as the following low-energy limit:

$$a_s = -\lim_{k \to 0} \frac{1}{k} \tan(\delta_0(k)) . \qquad (C.8)$$

In this way, one then finds

$$\lim_{k \to 0} \sigma(k) = 4\pi a_s^2 . \qquad (C.9)$$

Thus, for a very dilute and ultracold gas of atoms, the cross section $\sigma(k)$ of low-energy atoms can be indeed approximated as $\sigma(k) \simeq \sigma(0) = 4\pi a_s^2$, where the value of a_s must be obtained from the knowledge of s-wave scattering amplitude $f_0(k)$ or the s-wave phase shift $\delta_0(k)$.

All the described results crucially depend on the scattering amplitude $f(k,\theta)$. How can one explicitly calculate $f(k,\theta)$ for a given potential $V(\mathbf{r})$? Equation (C.1) admits formally the Lippmann-Schwinger solution:

$$\psi(\mathbf{r}) = e^{i\mathbf{k}\cdot\mathbf{r}} - \frac{m}{4\pi\hbar^2} \int d^3\mathbf{r}' \frac{e^{ik|\mathbf{r}-\mathbf{r}'|}}{|\mathbf{r}-\mathbf{r}'|} V(\mathbf{r}') \psi(\mathbf{r}') . \qquad (C.10)$$

In the far field, where

$$\frac{e^{ik|\mathbf{r}-\mathbf{r}'|}}{|\mathbf{r}-\mathbf{r}'|} \simeq \frac{e^{ikr}}{r}, \tag{C.11}$$

Eq. (C.10) becomes Eq. (C.2) with

$$f(k,\theta) = -\frac{m}{4\pi\hbar^2} \int d^3\mathbf{r}' \, V(\mathbf{r}') \, \psi(\mathbf{r}') . \tag{C.12}$$

Substituting in this formula $\psi(\mathbf{r}')$ with the incident plane wave $e^{i\mathbf{k}\cdot\mathbf{r}'}$, we obtain the s-wave Born approximation:

$$f_0(k) = -\frac{m}{4\pi\hbar^2} \int d^3\mathbf{r}' \, V(\mathbf{r}') \, e^{i\mathbf{k}\cdot\mathbf{r}'} = -\frac{m}{4\pi\hbar^2} \tilde{V}(k) , \tag{C.13}$$

for the s-wave scattering amplitude, with $\tilde{V}(k)$ the Fourier transform of $V(\mathbf{r})$. Working in the Born appoximation and considering the contact potential in Eq. (1.117), we find

$$a_s = \frac{m}{4\pi\hbar^2} g , \tag{C.14}$$

that is an explicit relation between the s-wave scattering length a_s and the strength g of the contact interaction. Considering instead the attractive square-well potential

$$V(\mathbf{r}) = \begin{cases} -V_0 & \text{for } r < R_0 \\ 0 & \text{for } r > R_0 \end{cases} \tag{C.15}$$

it is possible to derive the expression:

$$a_s = R_0 \left(1 - \frac{\tan(k_0 R_0)}{k_0 R_0}\right) \tag{C.16}$$

with $k_0 = \sqrt{mV_0/\hbar^2}$. The behavior of the scattering length a_s as a function of $k_0 R_0$ is quite remarkable: for $k_0 R_0 \to 0^+$, one has $a_s \to 0^-$, for $k_0 R_0 \to (\pi/2)^\mp$ follows that $a_s \to \pm\infty$. A divergent scattering length a_s means the formation of a bound state (also the scattering cross section diverges) and it is said that a resonance occurs. Actually, by increasing the potential depth V_0, one produces a new bound state and a resonance every time that $k_0 R_0 = (2j+1)\pi/2$ with j a natural number. This is the mechanism associated to the so-called Feshbach resonances, where the scattering length a_s diverges due to a change in the interaction potential induced by an external constant magnetic field $\mathbf{B}$. When $k_0 R_0 = j\pi$ the scattering length a_s vanishes identically, the scattering cross section is zero, and the target becomes invisible

(Ramsauer-Townsend effect). Neutral alkali atoms interact through short-range van der Waals interaction $V(\mathbf{r})$ far from resonance in ultracold atomic gases, which create an environment for the exploitation of resonant scattering processes. Indeed, it is possible to adjust the effective strength of interaction by permitting particles to create a virtual bound state known as a resonance. As previously said, a Feshbach resonance allows the two-atom system to be tuned via an external magnetic field **B**. In this way, the scattering length a_s can be modified from repulsive to attractive simply by changing the field **B**.

Scattering theory can be also formulated using the transfer matrix (T-matrix) formalism, where the starting point is the Hamiltonian operator

$$\hat{H} = \hat{H}_0 + \hat{V} , \tag{C.17}$$

with $\hat{H}_0 = \hat{\mathbf{p}}^2/(2m_r)$ the kinetic energy operator of a particle of reduced mass $m_r = m/2$ and linear momentum $\hat{\mathbf{p}}$, while $\hat{V}$ is the interaction potential operator such that $\hat{V}|\mathbf{r}\rangle = V(\mathbf{r})|\mathbf{r}\rangle$, where $|\mathbf{r}\rangle$ is the eigenstate of the position operator $\hat{\mathbf{r}}$. We introduce $|\mathbf{k}\rangle$ as the initial state of the problem, which is assumed to be eigenstate of the linear momentum operator $\hat{\mathbf{p}}$, i.e., $\hat{\mathbf{p}}|\mathbf{k}\rangle = \hbar\mathbf{k}|\mathbf{k}\rangle$, and it is such that

$$\hat{H}_0|\mathbf{k}\rangle = E_k|\mathbf{k}\rangle , \tag{C.18}$$

with $E_k = \hbar^2 k^2/(2m_r)$. Within this Dirac notation, remember that $\psi(\mathbf{r}) = \langle\mathbf{r}|\psi\rangle$, Eq. (C.1) is rewritten as

$$\left(\hat{H}_0 + \hat{V}\right)|\psi\rangle = E_k|\psi\rangle , \tag{C.19}$$

where $|\psi\rangle$ is the state we are looking for. The T-matrix is defined as the operator $\hat{T}$ such that

$$\hat{T}|\mathbf{k}\rangle = \hat{V}|\psi\rangle , \tag{C.20}$$

and its connection with the scattering amplitude $f(k, \theta)$ is given by the formula:

$$f(k, \theta) = -\frac{m}{4\pi\hbar^2} T_{\mathbf{k}\mathbf{k}'} , \tag{C.21}$$

where

$$T_{\mathbf{k}\mathbf{k}'} = \langle\mathbf{k}|\hat{T}|\mathbf{k}'\rangle \tag{C.22}$$

is a matrix element of $\hat{T}$ with the constraint that $|\mathbf{k}| = |\mathbf{k}'| = k$. After some manipulations of the previous equations, one finds that the matrix element $T_{\mathbf{k}\mathbf{k}'}$ satisfies the T-matrix equation:

$$T_{\mathbf{k}\mathbf{k}'} = V_{\mathbf{k}\mathbf{k}'} + \frac{1}{L^3} \sum_{\mathbf{q}} \frac{V_{\mathbf{k}\mathbf{q}}}{\frac{\hbar^2 k^2}{2m_r} - \frac{\hbar^2 q^2}{2m_r} + i0^+} T_{\mathbf{q}\mathbf{k}'}, \qquad (C.23)$$

where $|\mathbf{q}\rangle$ is an intermediate state and $V_{\mathbf{k}\mathbf{k}'} = \langle \mathbf{k}|\hat{V}|\mathbf{k}'\rangle = \tilde{V}(\mathbf{k} - \mathbf{k}')$.

In the special case of contact interaction, Eq. (1.117), the T-matrix equation gives

$$T_{\mathbf{k}\mathbf{k}'} = g + g \frac{1}{L^3} \sum_{\mathbf{q}} \frac{1}{\frac{\hbar^2 k^2}{2m_r} - \frac{\hbar^2 q^2}{2m_r} + i0^+} T_{\mathbf{q}\mathbf{k}'}. \qquad (C.24)$$

Perturbatively, at the lowest order, we have

$$T_{\mathbf{k}\mathbf{k}'} = g \qquad (C.25)$$

that is the Born approximation. In addition, setting $T(E_k) = T_{\mathbf{k}\mathbf{k}'}$ and working in the s-wave regime where $T_{\mathbf{k}\mathbf{k}'} \simeq T_{\mathbf{q}\mathbf{k}'}$, we obtain

$$T(E_k) = g + g\, T(E_k) \frac{1}{L^3} \sum_{\mathbf{q}} \frac{1}{E_k - \frac{\hbar^2 q^2}{2m_r} + i0^+}, \qquad (C.26)$$

namely,

$$\frac{1}{T(E_k)} = \frac{1}{g} - \frac{1}{L^3} \sum_{\mathbf{q}} \frac{1}{E_k - \frac{\hbar^2 q^2}{2m_r} + i0^+}. \qquad (C.27)$$

Finally, identifying $T(0)$ as the renormalized interaction strength g_r, we have

$$\frac{1}{g_r} = \frac{1}{g} + \frac{1}{L^3} \sum_{\mathbf{q}} \frac{m}{\hbar^2 q^2}, \qquad (C.28)$$

which is meaningful across a Feshbach resonance provided the inclusion of an ultraviolet cutoff Λ for the wave number $q = |\mathbf{q}|$.

Qubits and Entanglement

D

In this appendix, we briefly present some basic ideas of quantum information. We define single and multiple qubits and then discuss quantum entanglement, introducing the von Neumann and entanglement entropy.

The qubit is the unit of quantum information, namely, a two-level quantum state. Quantum computation uses the qubits to store information, and, by implementing operations involving the state of multiple qubits, the information can be manipulated in order to solve computational problems. There are many examples of two-level systems, for instance:

- {Horizontal, vertical} polarizations in light
- {Ground, excited} eigenstates in atom or nucleus
- {Left, right} wells in a double-well potential
- {Up, down} spins in a particle

The two (orthonormal) basis states to describe a qubit are usually denoted by $|0\rangle$ and $|1\rangle$ in analogy to the classical case of two bits 0 and 1. Unlike the classical case, a generic qubit is a linear superposition of these two basis states:

$$|\psi\rangle = \alpha|0\rangle + \beta|1\rangle, \tag{D.1}$$

where α and β are complex probability amplitudes such that $|\alpha|^2 + |\beta|^2 = 1$.

The two basis states of a qubit can be expressed using different choices. For example, we can write

$$|0\rangle = \begin{pmatrix} 1 \\ 0 \end{pmatrix}, \quad |1\rangle = \begin{pmatrix} 0 \\ 1 \end{pmatrix}, \tag{D.2}$$

by using a $\mathbb{C}^2 = \mathbb{C} \times \mathbb{C}$ representation. Another choice is, for instance,

$$\begin{cases} |+\rangle = \frac{1}{\sqrt{2}}(|0\rangle + |1\rangle) = \frac{1}{\sqrt{2}} \begin{pmatrix} 1 \\ 1 \end{pmatrix}, \\ |-\rangle = \frac{1}{\sqrt{2}}(|0\rangle - |1\rangle) = \frac{1}{\sqrt{2}} \begin{pmatrix} 1 \\ -1 \end{pmatrix}. \end{cases}$$

Another useful basis is

$$\begin{cases} |+i\rangle = \frac{1}{\sqrt{2}}(|0\rangle + i|1\rangle) = \frac{1}{\sqrt{2}} \begin{pmatrix} 1 \\ i \end{pmatrix}, \\ |-i\rangle = \frac{1}{\sqrt{2}}(|0\rangle - i|1\rangle) = \frac{1}{\sqrt{2}} \begin{pmatrix} 1 \\ -i \end{pmatrix}. \end{cases} \quad \text{(D.3)}$$

An important example of the usage of this last basis vectors is in quantum optics. Given the basis states $|H\rangle$ and $|V\rangle$ of horizontal and vertical polarization of light, often one uses

$$\begin{cases} |CW\rangle = \frac{1}{\sqrt{2}}(|H\rangle + i|V\rangle), \\ |CCW\rangle = \frac{1}{\sqrt{2}}(|H\rangle - i|V\rangle), \end{cases}$$

where CW stands for clockwise and CCW for counterclockwise. These states represent circularly polarized light with right- and left-handed polarization, related to the clockwise and counterclockwise states by the propagation direction.

From a particular representation of the basis vectors, one can compute the projector operators relative to the basis set, for instance, in the $\mathbb{C}^2 = \mathbb{C} \times \mathbb{C}$ representation:

$$|0\rangle\langle 0| = \begin{pmatrix} 1 & 0 \\ 0 & 0 \end{pmatrix}, \quad |0\rangle\langle 1| = \begin{pmatrix} 0 & 1 \\ 0 & 0 \end{pmatrix},$$

$$|1\rangle\langle 0| = \begin{pmatrix} 0 & 0 \\ 1 & 0 \end{pmatrix}, \quad |1\rangle\langle 1| = \begin{pmatrix} 0 & 0 \\ 0 & 1 \end{pmatrix}.$$

Similar expressions can be made for the other basis sets. We remark that the choice of the basis should be made by keeping in mind some properties of the specific problem one wants to solve. The chosen basis, if the aim is to perform computations, is often called computational basis.

D Qubits and Entanglement

The operations on a single qubit can be naturally generalized to multiple qubits. This is the case of interest for quantum computing: by tailoring the interactions between many two-level systems, we can design a quantum logic circuit in which we build algorithms composed of elementary instructions. A multiple N-qubit is a quantum state characterized by

$$|\Phi_N\rangle = \sum_{\substack{\{\alpha_1\alpha_1...\alpha_N\} \\ =\{0,1\}}} A_{\alpha_1\alpha_2...\alpha_N} |\alpha_1\rangle \otimes |\alpha_2\rangle \otimes ... \otimes |\alpha_N\rangle, \tag{D.4}$$

where $|\alpha_i\rangle$ $i = 1, ..., N$ and $\alpha_i = 0, 1$ is a single qubit. Thus, the N-qubit describes N two-state systems. Often, a multiple N-qubit is also called "quantum register" because it can be considered the quantum mechanical analog of a classical processor register.

We now focus on the case of a 2-qubit, for which we can define the most important gates. The most general 2-qubit is given by

$$|\Phi_2\rangle = A_{00}|0\rangle \otimes |0\rangle + A_{01}|0\rangle \otimes |1\rangle + A_{10}|1\rangle \otimes |0\rangle + A_{11}|1\rangle \otimes |1\rangle, \tag{D.5}$$

to simplify the notation, we use

$$\begin{aligned} |00\rangle &= |0\rangle \otimes |0\rangle \\ |01\rangle &= |0\rangle \otimes |1\rangle \\ |10\rangle &= |1\rangle \otimes |0\rangle \\ |11\rangle &= |1\rangle \otimes |1\rangle, \end{aligned} \tag{D.6}$$

in this way, we can write $|\Psi_2\rangle$ as

$$|\Phi_2\rangle = A_{00}|00\rangle + A_{01}|01\rangle + A_{10}|10\rangle + A_{11}|11\rangle. \tag{D.7}$$

We can introduce a basis state for 2-qubits analogously to the case of the single qubit. This is indeed the natural basis of the tensor product state:

$$|00\rangle = \begin{pmatrix} 1 \\ 0 \\ 0 \\ 0 \end{pmatrix}, \quad |01\rangle = \begin{pmatrix} 0 \\ 1 \\ 0 \\ 0 \end{pmatrix}, \quad |10\rangle = \begin{pmatrix} 0 \\ 0 \\ 1 \\ 0 \end{pmatrix}, \quad |11\rangle = \begin{pmatrix} 0 \\ 0 \\ 0 \\ 1 \end{pmatrix}, \tag{D.8}$$

which form a basis of a generic 2-qubit. Among the quantum gates acting on a 2-qubit, there are two relevant ones defined in this basis as:

- Identity gate

$$\hat{\mathbf{1}} = \begin{pmatrix} 1 & 0 & 0 & 0 \\ 0 & 1 & 0 & 0 \\ 0 & 0 & 1 & 0 \\ 0 & 0 & 0 & 1 \end{pmatrix} \qquad \begin{array}{l} |00\rangle \to |00\rangle \\ |01\rangle \to |01\rangle \\ |10\rangle \to |10\rangle \\ |11\rangle \to |11\rangle \,, \end{array} \tag{D.9}$$

- CNOT gate

$$\hat{C} = \begin{pmatrix} 1 & 0 & 0 & 0 \\ 0 & 1 & 0 & 0 \\ 0 & 0 & 0 & 1 \\ 0 & 0 & 1 & 0 \end{pmatrix} \qquad \begin{array}{l} \left. \begin{array}{l} |00\rangle \to |00\rangle \\ |01\rangle \to |01\rangle \end{array} \right\} |\beta\rangle \text{ remains equal,} \\ \left. \begin{array}{l} |10\rangle \to |11\rangle \\ |11\rangle \to |10\rangle \end{array} \right\} |\beta\rangle \text{ changes.} \end{array} \tag{D.10}$$

The CNOT gate can be understood simply in terms of the so-called target and control qubits: given $|\alpha\beta\rangle = |\alpha\rangle \otimes |\beta\rangle$, the target qubit $|\beta\rangle$ changes only if the control $|\alpha\rangle$ qubit is set to $|1\rangle$; otherwise, it stays the same.

The 2-qubit state $|\Phi_2\rangle$ is called separable if it can be written as the tensor product of two generic single qubits $|\Psi_A\rangle$ and $|\Psi_B\rangle$, i.e.,

$$|\Phi_2\rangle = |\Psi_A\rangle \otimes |\Psi_B\rangle . \tag{D.11}$$

If this is not possible, the state $|\Phi_2\rangle$ is called entangled. Let us be concrete with this simple example: the simplest 2-qubit state

$$|\Phi_2\rangle = A_{01}|01\rangle + A_{10}|10\rangle, \tag{D.12}$$

is separable only if $A_{01} = 0$ or $A_{10} = 0$. In fact, let us set

$$|\psi_A\rangle = \alpha_A|0\rangle + \beta_A|1\rangle, \quad |\psi_B\rangle = \alpha_B|0\rangle + \beta_B|1\rangle \tag{D.13}$$

A necessary condition for the state $|\Phi_2\rangle$ to be written as $|\psi_A\rangle \otimes |\psi_B\rangle$ is

$$\begin{cases} \alpha_A \alpha_B = 0 \\ \beta_A \beta_B = 0 \end{cases} \tag{D.14}$$

these equations can be satisfied only if at least one of the coefficients is zero. Since $A_{01} = \alpha_A \beta_B$ and $A_{10} = \beta_A \alpha_B$, we conclude that at least one of this two coefficients is null.

D Qubits and Entanglement

From our previous discussion, it is clear that

$$\frac{1}{\sqrt{2}}(|01\rangle \pm |10\rangle) = |\psi^\pm\rangle, \text{ but also } \frac{1}{\sqrt{2}}(|00\rangle \pm |11\rangle) = |\Phi^\pm\rangle, \quad (D.15)$$

are entangled states. Instead, examples of separable states are

$$|01\rangle, \text{ and } |00\rangle, \text{ as well as } \frac{1}{\sqrt{2}}(|01\rangle + |11\rangle), \quad (D.16)$$

because the latter can be expressed as

$$\frac{1}{\sqrt{2}}(|0\rangle + |1\rangle) \otimes |1\rangle. \quad (D.17)$$

Entangled states are extremely important in quantum information. The four states in Eq. (D.15) are also called Bell states. When measuring a single qubit on an entangled state, the measurement will determine the state of the other qubit. Before the measurement, in the other hand, neither of the qubits has a definite property; they are uncommitted.

While the entanglement property of a given state is categorical, namely, we can have it or not, for entangled states, we can define a quantity that allows one to quantify the entanglement. The quantity that we now introduce, namely, the von Neumann entropy, can be used as such a value for pure states. In general, there are many measures of entanglement, and it is not possible to use a unique one for classifying entanglement of the states, yet they are useful as analytical tools. To measure the entanglement of a pure state

$$|\psi\rangle_{AB} \in \mathcal{H}_A \otimes \mathcal{H}_B, \quad (D.18)$$

the first step is to write its density matrix given by

$$\hat{\rho}_{AB} = |\psi\rangle_{AB} \, {}_{AB}\langle\psi|. \quad (D.19)$$

The second step is to get the partial density matrices:

$$\hat{\rho}_A = \text{Tr}_B \left[\hat{\rho}_{AB}\right],$$
$$\hat{\rho}_B = \text{Tr}_A \left[\hat{\rho}_{AB}\right].$$

The third, and final, step is to calculate the entanglement entropy $\varepsilon(\hat{\rho}_{AB})$ defined as

$$\varepsilon\left(\hat{\rho}_{AB}\right) = S\left(\hat{\rho}_A\right) = S\left(\hat{\rho}_B\right), \quad (D.20)$$

where $S(\hat{\rho}_A)$ and $S(\hat{\rho}_A)$ are the von Neumann entropies of the partial densities $\hat{\rho}_A$ and $\hat{\rho}_B$, namely,

$$S(\hat{\rho}_A) = -\text{Tr}\left[\hat{\rho}_A \log_2(\hat{\rho}_A)\right] \tag{D.21}$$

$$S(\hat{\rho}_B) = -\text{Tr}\left[\hat{\rho}_B \log_2(\hat{\rho}_B)\right]. \tag{D.22}$$

In general, for a separable state, one finds $\varepsilon(\hat{\rho}_{AB}) = 0$, while an entangled state has instead $\varepsilon(\hat{\rho}_{AB}) > 0$. In the special case of a maximally entangled state, it is possible to prove that $\varepsilon(\hat{\rho}_{AB}) = \log_2 N$, with $N = \dim(\mathcal{H}_A)$. This is the case of Bell states.

Let us consider $|\Psi\rangle_{AB} \in \mathcal{H}_A \otimes \mathcal{H}_B = \mathcal{H}$. Its density matrix is

$$\hat{\rho}_{AB} = |\psi\rangle_{AB}\, {}_{AB}\langle\psi|. \tag{D.23}$$

The density matrix $\hat{\rho}_{AB}$ of the "subsystem A" of the Hilbert space $\mathcal{H}_A$ is defined as

$$\hat{\rho}_A = \text{Tr}_B\left[\hat{\rho}_{AB}\right] = \sum_K {}_B\langle K|\hat{\rho}_{AB}|K\rangle_B, \tag{D.24}$$

namely, the partial trace with respect to the subsystem B, where $\{|K\rangle_B\}$ is a basis of $\mathcal{H}_B$. We remark that the subsystem of a pure state can be a mixed state. In fact, if we consider the pure (but entangled) state

$$|\psi\rangle_{AB} = \frac{1}{\sqrt{2}}(|0\rangle_A|0\rangle_B + |1\rangle_A|1\rangle_B) \tag{D.25}$$

it has density matrix:

$$\hat{\rho}_{AB} = |\psi\rangle_{AB}\, {}_{AB}\langle\psi|$$
$$= \frac{1}{2}\{|0\rangle_A|0\rangle_B\, {}_A\langle 0|\, {}_B\langle 0| + |0\rangle_A|0\rangle_B\, {}_A\langle 1|\, {}_B\langle 1| \tag{D.26}$$
$$+ |1\rangle_A|1\rangle_B\, {}_A\langle 0|\, {}_B\langle 0| + |1\rangle_A|1\rangle_B\, {}_A\langle 1|\, {}_B\langle 1|\},$$

and its subsystem A has density matrix:

$$\hat{\rho}_A = {}_B\langle 0|\hat{\rho}_{AB}|0\rangle_B + {}_B\langle 1|\hat{\rho}_{AB}|1\rangle_B = \frac{1}{2}|0\rangle_A\, {}_A\langle 0| + \frac{1}{2}|1\rangle_A\, {}_A\langle 1| \tag{D.27}$$

The von Neumann entropy for a quantum state described by the density matrix $\hat{\rho}$ is given by

$$S(\hat{\rho}) = -\text{Tr}\left[\hat{\rho}\log_2(\hat{\rho})\right]. \tag{D.28}$$

D Qubits and Entanglement

If $\{\lambda_i\}$ are the eigenvalues of $\hat{\rho}$, then

$$S(\hat{\rho}) = -\sum_i \lambda_i \log_2(\lambda_i). \tag{D.29}$$

To justify its introduction, we proceed now giving some examples.

1. For the state $|0\rangle$, we have $\hat{\rho} = |0\rangle\langle 0|$ whose eigenvalues are $\{1, 0\}$. It follows immediately that $S(\hat{\rho}) = 0$.
2. For the state $\sqrt{1/3}|0\rangle + \sqrt{2/3}|1\rangle$, we have

$$\hat{\rho} = \frac{1}{3}\begin{pmatrix} 1 & \sqrt{2} \\ \sqrt{2} & 2 \end{pmatrix} \tag{D.30}$$

   whose eigenvalues are again $\{1, 0\}$. It follows immediately that $S(\hat{\rho}) = 0$.
3. For the state described by $\hat{\rho} = \frac{1}{3}|0\rangle\langle 0| + \frac{2}{3}|1\rangle\langle 1| = \frac{1}{3}\begin{pmatrix} 1 & 0 \\ 0 & 2 \end{pmatrix}$, the eigenvalues of $\hat{\rho}$ are $\{\frac{1}{3}, \frac{2}{3}\}$. It follows that $S(\hat{\rho}) = -\frac{1}{3}\log_2\left(\frac{1}{3}\right) - \frac{2}{3}\log_2\left(\frac{2}{3}\right) = \log_2(3) - \frac{2}{3}$.

In general, one can prove that:

- $S(\hat{\rho}) \geqslant 0$.
- $S(\hat{\rho}) = 0$ only if $\hat{\rho}$ describes a pure state.
- The maximum value of $S(\hat{\rho})$ is $\log_2(N)$ with $N = \dim(\mathcal{H}_\ell)$.

Indeed, the Bell states in Eq. (D.15) achieve the maximum of the entanglement entropy for a 2-qubit system. They are also called maximally entangled states.

Printed in the United States
by Baker & Taylor Publisher Services